URBAN GOVERNANCE IN SOUTHERN EUROPE

Urban and Regional Planning and Development Series

Series Editor: Professor Graham Haughton

Based on over a decade of publishing the highest quality research, the Urban and Regional Planning and Development Series has developed a strong profile. It is internationally recognised for its high quality research monographs. The emphasis is on presenting original research findings which are informed by theoretical sophistication and methodological rigour. It is avowedly global in its outlook, with contributions welcomed from around the world. The series is open to contributions from a wide variety of disciplines, including planning, geography, sociology, political science, public administration and economics.

Other titles in the series

Strategic Planning for Contemporary Urban Regions
City of Cities: A Project for Milan
Alessandro Balducci, Valeria Fedeli and Gabriele Pasqui
ISBN 978 0 7546 7967 7

Inventive City-Regions
Path Dependence and Creative Knowledge Strategies
Marco Bontje, Sako Musterd and Peter Pelzer
ISBN 978 1 4094 1772 9

Discourse Dynamics in Participatory Planning
Opening the Bureaucracy to Strangers
Diana MacCallum
ISBN 978 0 7546 7296 8

Planning Cultures in Europe
Decoding Cultural Phenomena in Urban and Regional Planning
Edited by Joerg Knieling and Frank Othengrafen
ISBN 978 0 7546 7565 5

De-coding New Regionalism
Shifting Socio-political Contexts in Central Europe and Latin America
Edited by James W. Scott
ISBN 978 0 7546 7098 8

Urban Governance in Southern Europe

Edited by

JOÃO SEIXAS
Instituto de Ciências Sociais da Universidade de Lisboa, Portugal

and

ABEL ALBET
Universitat Autònoma de Barcelona, Spain

LONDON AND NEW YORK

First published 2012 by Ashgate Publishing

2 Park Square, Milton Park, Abingdon, Oxon OX14 4RN
711 Third Avenue, New York, NY 10017, USA

Routledge is an imprint of the Taylor & Francis Group, an informa business

First issued in paperback 2016

British Library Cataloguing in Publication Data
Urban Governance in Southern Europe.
– (Urban and Regional Planning and Development Series)
1. Municipal government – Europe, Southern. 2. Urban policy – Europe, Southern.
I. Series II. Seixas, Joao. III. Albet i Mas, Abel.
320.8'5'094–dc23

Library of Congress Cataloging-in-Publication Data
Seixas, Joao.
Urban Governance in Southern Europe / by Joao Seixas and Abel Albet.
p. cm. – (Urban and Regional Planning and Development Series)
Includes bibliographical references and index.
1. Municipal government – Europe, Southern. I. Albet, Abel. II. Title.
JS3000.S45 2013
320.8'5094–dc23 2012025196

ISBN 978-1-4094-4434-3 (hbk)
ISBN 978-1-138-26694-0 (pbk)

Contents

List of Figures, Maps and Tables

Figures

Maps

Tables

Notes on Contributors

Abel Albet is Associate Professor at the Department of Geography of the Universitat Autònoma de Barcelona (Spain). His teaching and research interests are focused on critical geographies (geographical thought; postmodern geographies), on urban studies (urban planning; urban governance; urban representations, discourses and images; public space; social problems of cities and metropolitan areas; Barcelona) and on new cultural geographies (urban cultural landscapes; orientalism and postcolonialism; travel accounts; Morocco). He is member of the steering committee of the International Critical Geography Group and editor of the book series *Espacios Críticos* (Icaria Editorial).

Irena Bačlija is an assistant professor at the Department of Political Science at the Faculty of Social Sciences (University of Ljubljana, Slovenia) and a research fellow at the Centre for the Analysis of the Administrative-Political Processes and Institutions. She lectures at the pre-graduate and post-graduate courses on the programme Policy Analyses and Public Administration. Her area of research is urban management, urban governance, local (self-)government and public administration reform. She has published many scientific articles, papers and books in the Slovenian and English languages.

João Cabral, Architect (Escola Superior de Belas Artes, Lisbon), Honours Diploma, Urban and Regional Development Planning (Architectural Association, London), PhD (Urban and Regional Studies, University of Sussex). He is professor at the Faculty of Architecture, Technical University of Lisbon (Portugal), and researcher at CIAUD – Research Centre for Architecture, Urbanism and Design. He has been involved in research projects on governance and policies in urban regions in Europe and in South and North America, comparing and evaluating spatial planning practices and urban regulation systems.

Ioannis Chorianopoulos is Assistant Professor at the Department of Geography, University of the Aegean (Greece). He has taught at the London School of Economics and the University of the Aegean, and has researched at the Hellenic Observatory (LSE). He is a member of the Royal Geographical Society, the Regional Studies Association and the LSE Urban Research Centre. His research interests and publications record focus on the themes of European integration, urban competitiveness and governance, spatial segregation, social exclusion and, recently, urban sprawl.

José Luís Crespo, geographer, holds the Masters Degree in Human, Regional and Urban Geography (University of Lisbon). He is lecturer at the Faculty of Architecture, Technical University of Lisbon (Portugal), and researcher at CIAUD – Research Centre for Architecture, Urbanism and Design. He is currently earning his PhD in Urban and Regional Planning at the Technical University of Lisbon on urban governance and planning in metropolitan areas.

Joaquin Farinós-Dasí is Regional Geographical Analysis Full Professor, responsible for the Territorial Governance, Spatial Planning and Environment research area at the Inter-University Institute for Local Development, University of Valencia (Spain). His work has been mainly focused on territorial and urban governance, spatial planning, European regional and cohesion policy, European institutions and policies, impact assessment, political regional geography, local and regional development, and international cooperation for development. He has experience in several research projects in Spain, the EU and Latin America, as coordinator, leading partner and/or partner. He is author of several publications on these topics in Spanish, English, French and Italian.

Francesca Governa, Master Degree in Architecture with a thesis on Urban and Regional Geography at the Politecnico di Torino, PhD in Territorial Planning and Real Estate Market, she is currently associate professor of Economic and Political Geography at the Politecnico di Torino (Italy) where she teaches Urban Geography. She carries out didactic and research activities in Italy and abroad. She has participated in research programmes at national and international level. Her main research activities concern urban geography, local development, and urban and territorial governance.

Christian Lefèvre has been working on issues of metropolitan government and governance in France and in Europe for the French ministry of Public Works, the City of Paris, the Region Ile-de-France, the European Commission, the European Science Foundation (ESF) and many other international (OECD, United Nations, Inter-American Development Bank) and foreign organisations and local governments (Italy, Germany, Denmark, the Netherlands, etc.). His major interests concern the issues of metropolitan public policies and politics, development strategies and the international relations of cities for which he was co-chair of the ESF research network, CITTA (2002–5). He is also member of the executive committee of the European Urban Research Association (EURA) and the Chief Editor of the on =line academic journal *Metropoles*. He has published several books and many articles on these issues. His most recent publications include: *Gouverner les métropoles* (2009); *Le strategie internazionali delle città* (with E. D'Albergo, 2007); and 'The democratic governance of metropolitan areas: international experiences and lessons for Latin American cities' in E. Rojas et al. *Governing the Metropolis* (2008).

Oriol Nel·lo is professor of Urban Geography at the Autonomous University of Barcelona (Spain), where among other tasks he served as Director of the Institute for Metropolitan Studies of Barcelona (1988–9). Also a politician, he was Member of the Catalan Parliament (1999–2004) and served from December 2003 to January 2011 as Secretary for Spatial Planning in the Autonomous Government of Catalonia, playing a leading role in land policy, neighbourhood rehabilitation and the long-awaited strategy for the vast metropolitan region around Barcelona.

Gilles Pinson is Professor of Political Science at Sciences-Po Lyon (University of Lyon, France). He is a researcher at TRIANGLE (UMR 5206, common laboratory with ENS-LSH, Sciences-Po Lyon, Université Lumière Lyon 2 and CNRS). He also teaches on the Master's Degree in 'Territorial and Urban Strategies' at Sciences-Po Paris. His research encompasses planning, urban policies and governance and the challenges of metropolitan cooperation. He has recently published *Gouverner par projet. Urbanisme et gouvernance des villes européennes* (2009).

Joan Romero-González is Human Geography Full Professor, former Director of the Inter-University Institute for Local Development at the University of Valencia (Spain) and Visiting Scholar at Leeds University School of Geography. He has recently focused his research and teaching activity on political geography, public policies, polity and new ways of territorial governance in Spain and Europe. He is coordinator of the Spanish research project 'Territorial Cooperation Strategies in Spain', co-founded by the Spanish Ministry of Education and Science and ERDF, and member of the Spanish LP team of ESPON Project 2.3.2 on Territorial and Urban Governance. Since 2002 he has authored more than 40 publications related with these topics.

João Seixas is research fellow at the Institute of Social Sciences of the University of Lisbon (Portugal), holds a PhD in Urban Geography from the Autonomous University of Barcelona and in Sociology of the Territory and the Environment from the ISCTE in Lisbon. He has been researcher and professor in Lisbon, Barcelona and Rio de Janeiro. He has coordinated several research projects (conceptual and applied) on contemporary tendencies in the political and social domains of cities and metropolises; on structures and dynamics of urban governance; on urban and territorial planning and development. He is also Commissioner of the Strategic Charter for the city of Lisbon. He has published several works in these areas, in different publications. He also exercises diverse civic activity in city governance, socioeconomic local development and cultural ambits.

Carla Tedesco, PhD in Planning and Regional Policy, and Assistant Professor at the IUAV University of Venice, where she teaches Urban Management. She carried out research projects on the influence of the EU on urban policy in Italy in a comparative perspective within the EU, being part of national and international groups. Her publications include: *Città in periferia* (Carocci, 2009 (co-author));

Rethinking European Spatial Policy as a Hologram. Actions, Institutions, Discourses (Ashgate, 2006 (co-editor)).

Nil Uzun is a City and Regional Planner. She earned her PhD in Urban Geography at Utrecht University in 2001. Currently, she holds the position of Associate Professor at the Department of City and Regional Planning at Middle East Technical University (Ankara, Turkey). Along with teaching urbanisation and urban sociology, she is doing research mainly on gentrification and urban residential transformation. She has published several articles in international journals and book chapters in national books.

Introduction[1]

João Seixas[2] and Abel Albet[3]

L'ordre routier et urbain est, par excellence, l'ordre humain de la Méditerranée.
Fernand Braudel (1949)

Cities of the South

The Mediterranean is one of the richest and most complex regions of the world. A richness and complexity erected throughout a deep and long established cultural and socio-economic history, and intrinsically connected with the evolution of its cities.

In this book we will open and in many ways synthesise several points for reflection concerning the present governance tendencies, possibilities and dilemmas facing the urban territories and societies of Southern Europe. This focus is supported on several bases. On one hand, bringing an overview on today's most pressing city policy developments and debates – namely those concerning urban socio-political (re)positioning tendencies in the face of profound social and economic transformations. On the other hand, by following some reference works developed upon the differences or even uniqueness of the realms of the Southern European city, on its urban landscapes and on its socio-political pulses. Finally, in the 11 chapters here included we also analyse several specific realities posed to renowned Southern European urban regions, ranging from Lisbon to Istanbul.

There are other criteria included here. We focus on the urban nodes and systems of the Northern shore of the vast territory that Braudel (1949) named the 'true Mediterranean' – the Southern European socio-cultural landscapes geographically and historically defined by the limits of the olive and fig trees. For centuries the two sides of the Euro-Mediterranean world – the Ottoman and Orthodox east and the Catholic west – gave rise to deep religious and political clashes. But the effects of recent and strong influential elements such as the political democratisation processes, the European Union, and remarkably newer cosmopolitan social and cultural tendencies, quite visible in urban daily life, in social movements and civic

---

1 An early version of this text was published in 2010 in *Análise Social. Revista do Instituto de Ciências Sociais da Universidade de Lisboa*, 197, 611–20.

2 Instituto de Ciências Sociais; Universidade de Lisboa. Av. Professor Aníbal de Bettencourt, 9; 1600-189 Lisbon (Portugal). jseixas@ics.ul.pt.

3 Universitat Autònoma de Barcelona; Departament de Geografia. 08193 Bellaterra (Spain). abel.albet@uab.cat.

expressions – all lead most of the Southern European urban world from a peripheral to a semi-peripheral, if not central, condition (Leontidou 1990, 2010). This has reduced the importance of religious and political regional differences. It is an evolution, however, that in quite an interesting manner is probably permitting the beginning of a better comprehension of the importance of common Mediterranean socio-cultural structures, and that has effectively set a different pace for capitalist transformation and development itself – and most surely, a different pace for its own urban growth and urban condition.

Southern European cities and most notably the largest metropolitan regions might configure themselves not only through some common specific geographical and morphological elements and formations, such as the combination of resourceful and compact urban centres with vast rural hinterlands – albeit these nowadays mostly urbanised and configuring strongly diffused metropolitan regions – but also through concrete forms of social and cultural expressions that have been configuring quite complex urban social humus. A cultural complexity erected in long-standing dense cities but presently sprawled urban structures, through highly spontaneous and intricate forms of social interactions and relationships, mixing both local and international networks (due both to historical immigration as well as to more recent emigration patterns), and where similarly complex urban *genius loci* have developed (Pace 2002, Coletta 2008). As Leontidou (1990: 2–3) wrote:

> The most striking similarities among Southern cities mostly stem from the coexistence of 'modernity' and informality (not 'tradition') on many levels as their class structure approaches the pattern of late capitalism, self-employment remains widespread; managers and executives coexist with artisans, shopkeepers and free labourers; in the location of economic activity, as CBD are rebuilt with modern office blocks, missed land uses predominate; in housing allocation, as modern apartment blocks spring up, self-built neighbourhoods continue to mushroom; in urban development, several private and public, customary and irregular (illegal and informal) strategies coexist and affect the systems of production and reproduction.

Certainly, and as Braudel (1949) so well expressed, the Mediterranean complexity is composed of a continuous superimposition of civilisations, several differential layers that always defied or even refuted any idea of unification or common characterisation. In fact, in not a few epochs it looks as if Mediterranean history and politics have been making their own way relatively apart from cultures and societies. This is a seemingly strange but above all intricate socio-cultural harmony, in our belief mainly projected by the lights and shadows expressed by the Southern *polis*.

These perspectives point to the understanding of the city as a vast world of social, economic and cultural interactions, thus as a local society with its own ecological dynamics and processes (Bagnasco and Le Galès 2000). Likewise, it is relevant to focus upon some main conceptual (and steadily applied) socio-political debates, such as the ones around the cultural and relational notions of social capital and cultural capital (Bourdieu 1997, Coleman 1990, Putnam 1993) and obviously, on the notion of governance (Jessop 1998, Jouve 2003, Maloney et al. 2000). Relatively different social and cultural urban scenarios shape relatively different socio-political structures and correspondingly different governance networks and actors' stakeholding. It might therefore be the case for the discussion of Mediterranean forms of urban governance, with its specific cultures, communities, processes, achievements – and dilemmas.

There are several common and specific socio-political tendencies shaping Euro-Mediterranean urban societies. And today, with a recognisably growing role being placed on cities as actors for human and sustainable development, it also shows that the ways these urban societies might face their common and specific future — as better or worse organised and strategy-supported collective actors — will certainly shape the future of the entire Mediterranean region itself, and a quite significant part of the future of Europe as a whole. Cities were always the main actors shaping the evolution of the Mediterranean world, as well as the whole of the European world. Empires were built from Lisbon and Madrid; spiritual expansions made from Rome and Córdoba; mercantile worlds erected from Venice and Seville; strong cultural lights shone from Athens and Istanbul. More recently, specific forms of industrialism, modernism, post-modernism have been sustained in the milieus of Milan, Lyon and Barcelona; symbolic battles for political liberty and autonomy were fought in cities like Paris and Sarajevo. All in all, these are historical colourful pictures that reveal both an absolute diversity and a systemic developmental nature in a continuously socio-political and cultural construction, strongly owing to city pulses, certainly continuing today in an ever-growing complex and polycentric world.

It is quite clear that these cities are today facing the rise and the consequences of considerable new urban as well as economic pressures. However, for many observers it does not look as evident that their corresponding civic societies, institutional and political systems and governance networks are sustaining the socio-cultural and political capabilities to adequately accompany the pace of change and demand. The most pressing challenges, such as several types of economic imbalances, social inequality, indebtedness, world immigration, peak-oil horizons, financial and real-estate crashes – along with the steady rise of new forms of cultural, labour and family arrangements, different forms of political cultures and citizenship expressions (Clark and Hoffman-Martinot 1998, Leontidou 2010), certainly configure high doses of both the opening of new socio-political possibilities as well as of vast spatial and political uncertainty.

The last couple of decades have seen the emergence of several political proposals to tackle some of these wide urban challenges, from state restructuring to new types of urban policies, governance and participation structures – and maybe urban politics itself. But serious doubts remain – more and more with the rise of the most recent and extremely pandemic global economic and institutional crisis, surely withdrawing resources from several local spheres – whether or not these most relevant political tendencies might be able to muster capacities to make collective enforcements through the thousands of local, partisan and considerably weak meridional forms of subsidiarity.

In other perspectives, new questions also arise with the restructuring of the semi-organic meridional governance forms of urban regimes. This might be the case if the new types of empowered urban configurations and political communities approach those that Brenner (2004) characterised and named as *Glocalizing Competition State Regimes*, and whose practices accord to what Mozzicafreddo et al. (2003) described as forms of *institutional particularism*. These are new types of sub-national governmentally coordinated and highly project-driven 'state-spaces', semi-democratic sub-regimes with strong control from corporate elites and specific political communities whose focuses and strategies are arranged around important urban resources, such as highly symbolic large projects and high yield returning real-estate investments.

However, at the same time evidence is also showing that the growing political role of cities is coming from paradigmatic shifts upon socio-cultural cognition and expression of urban societies. Namely in the development of much more cosmopolitan forms of education, of social and spatial identification, and of concomitant expressions of urban activity, labour and political attitudes. Strong cultural tendencies that are surely opening new dimensions of urban civic exigency, of governance stakeholding and dynamics, and of democracy itself.

The combination of the long Mediterranean history with the most recent but enduring global processes might show to us that the future of Southern European cities will certainly have to depend – and thus to interpret – upon its own socio-cultural networks, dynamics and cosmopolitan evolution. Thus giving a central role to such relational and cultural dimensions as the ones concerning urban governance configurations, urban regimes, socio-political conflicts and strategic choices.

Meanwhile, new and dramatic variables arise. The effects of the present economic global crisis have had a heavy impact on Southern European cities, societies and economies. The contraction of public and private investment – or, directly, the public and private disinvestment – on welfare state structures and facilities, dangerously exacerbate spatial inequalities and social injustices. Southern city politics seems now growingly marked by corporate relocation, skilled emigration, capital flight, increased tourism and volatile investments. Cities, as political actors, are again being left on their own in a context almost solely supported by 'the nature of markets'. Major changes taking place in European cities seem designed in order to legitimise actions and interests of neoliberal perspective (Jessop 2002), too often reducing urban strategic thinking and even urban planning to discourse generalisations and discredit.

For the moment, there is an evident triumph of the financial economy hegemonic discourse as the essential reference for overall politics. In the framework of neoliberalism, financial economy itself became a form of proto-ideology (Massey 2011). All urban space, including public urban space, is being seen as a capital resource, thus subject to its circuits, actors and valuations. The wide dissemination of the 'no-alternative' discourse pressures to the depoliticisation of urban thought itself, including some of its most fundamental areas of public service like the tackling of socio-spatial inequalities or the sustained development of urban quality of life (Hall and Massey 2010). Increasingly, the city is strangely less seen as a habitat for human progress and endeavour, but again and mainly as a competitive and entrepreneurial arena. It might be thus the case that, as Brawley (2009) states, 'cities are increasingly less powerful than they look'. This has most probably meant a radical reconsideration of urban government and urban governance schemes. Simultaneously, and corresponding to a growing cosmopolitanism, contra-pressure comes from citizens. The occupation of streets and squares, both at central or peripheral places of the Southern European metropolis by growing populational fringes, is an unequivocal demonstration of the perversions of present mainstream politics, and a growingly convincing call for renovated structures and dynamics of urban governance (Mayer 2007).

South European Urban Governance

The 11 chapters presented in this book are quite varied, both in their empirical objects of analysis as well in their analytical and critical positioning. They reflect and expand in very stimulating manners the overall reflections above expressed.

Based on empirical evidence taken in several Southern European cities, Christian Lefèvre does not focus his text on a single urban region or specific process or policy. Nonetheless, his analysis of the difficulties in the production of new urban political spaces that might better configure and tackle the scale of main urban challenges, fits extremely well into the bulk of the dilemmas presently facing most of the Mediterranean cities. Focusing his attention on the scale of the metropolis, he states that this most relevant urban functional scale has not been the focus of the most enduring and highlighted processes such as the political decentralisation ones in France and Italy, or the steady empowerment of local societies in many other urban territories. It is considerably ironic to realise that the only big city to have an effective metropolitan government of its own right – Madrid – had it established by chance. This means that most of the proposals of institutional transformation and reform tendencies of the last decades did not, after all, effectively manage to configure the city or its urban realities and needs, as the main objective of change. There are, nonetheless, some processes such as the ones of Barcelona or Turin, where the main purpose has been to slowly build political thickness at the metropolitan scale through governance procedural and sectorial arrangements co-opted by different urban institutions (public and private) in each metropolis. At the same time, the local and neighbourhood political empowerment processes followed in various cities might also provide support conditions for the wider reconfiguration of the political city. Some authors refer to these governance constructions (developed through the most varied projects, although not so many under a global strategy) as a sort of *variable geography*, which by its flexibility to the formation of sub-collective choices might better raise new kinds of identity, namely at the metropolitan scale, and thus achieve the new political reinforced spaces that are desired. But Lefèvre remains relatively cautious about these possibilities, stating not only the risks of partial governance empowerment processes being mainly built on opposition to broader collective objectives, but remembering as well the fragilities themselves of socio-political edifications mostly sustained by governance networks.

The chapter of José Luis Crespo and João Cabral follows these preoccupations. These authors expose the paradoxical situation of the Lisbon metropolitan area – namely in its dependency on the important but highly uncertain urban governance future developments. In face of a nonexistent effective metropolitan political institution and of ever peripheral and incomplete urban planning capabilities – contributing decisively to the fragmentation not only of many socio-economic metropolitan dynamics, but also to the fragmentation of several political responsibilities – the authors also emphasise the role of governance in the possible shape of the future metropolis. Here, even the strongest planning instruments – the municipal plans – are themselves a reflection of considerable local, albeit quite diverse, compromises. In a metropolitan area like Lisbon, if the most recent regional strategies and policies show to be almost absolutely dependent on the materialisation of cooperation networks, it is still also evident that stronger political or institutional enforcements might prove to be rather difficult vis-à-vis powerful interests ranging from the vast and complex public and municipal universes to the most varied private or single-project oriented strategies. Within this focus, therefore, they warn of the uncertainties and weaknesses that remain – more than limitations – regarding urban governance enhancement and capabilities, stating that 'governance definitely opened up a field of research that is far from being explored'.

Francesca Governa states that the emergence and consolidation of new means of governance is not so much a consequence of the decline of the power of the state and public administration, but rather of their (in)capacity to adapt to external and internal changes. To substantiate such changes, Governa begins her chapter with a broad review of the transformations observed in the political and institutional ordination of the Italian public administrations, and especially the consequences of several changes in the territorial and urban management, well evidenced through governance tendencies and dilemmas, and the growing orientation to see urban issues mostly through symbolic and populist themes, likes those surrounding security, immigration and mega-events. Likewise, Governa quite critically reveals the inadequacy of the institutional changes operated in Italy and the impossibility to speak of a genuine urban policy presently being followed in her country. In terms of public strategies, mainly discursive, all of the changes have revolved around two seemingly intrinsically incompatible concepts – competitiveness and cohesion – which are nevertheless presented as simultaneous and interdependent objectives in the Italian urban policies.

Carla Tedesco discusses urban governance in Southern Italy from a twofold perspective: on the one hand it overcomes the traditional analytical framework of the large geographical division between the richer North-Centre and the poorer South. Southern Italian cities are not analysed as homogeneous entities: the interactions between urban actors are related to the diversity of the different sub-regions as well as framed within the Europeanisation processes. On the other hand, the text focuses on a quite interesting issue, a common problem dealt within most of these Italian urban societies: the difficulties of producing effective outcomes by most political and policy innovation efforts, deeply marked by both traditional organisational and institutional weaknesses and everyday manifestations of informal status quo practices.

Oriol Nel·lo's considerably differential text shows in a very stimulating manner the relatively specific paths of some of Catalonia's urban policies. He presents the details (objectives, results and limitations) of the 'Law on Neighbourhoods' approved and implemented annually between 2004 and 2010 by the Autonomous Government of Catalonia. This Law is in itself a fairly atypical but extremely positive example of the fact that it is possible to implement quite interesting formulas of institutional governance through effective urban policies based on agreements between different public administrations (even governed by divergent or clearly opposed political parties). The exceptional nature of this case also lies in its dimensions (the high number of people and neighbourhoods that benefit from it and the total investments committed) and its characteristics (with an emphasis on transversal projects able to boost new resources and action; acting on neighbourhoods of medium-sized towns in semi-rural surroundings as much as in metropolitan areas). The tacit and coherent distribution of functions between regional and municipal administrations, and the involvement of residents in the entire process appear to be the potential keys to the success of the experiment, reflected in the effective transformation of these neighbourhoods and the prevention of a number of prospective problems. On the basis of the process initiated through this Law, the author proposes 'ten lessons' as an example of 'good practice' for participative, proactive and effective governance.

Joan Romero and Joaquín Farinós explore the failures to develop new metropolitan governance practices of most Spanish urban territories. They report structural problems when joint projects are developed; denouncing constant cross-sectoral conflicts at inter-institutional multi-level, as well as showing the fissures of a possible territorial cooperation culture, between classical administrative fragmentation. There are therefore strong barriers to the support of federally inspired democratic urban multi-level state governance experiences, whose complexities require well developed coordination and cooperation routines and complicities. These authors analyse the causes and the current initiatives of spatial planning and territorial cooperation at the metropolitan scales, proposing possible debatable agendas for the near future.

Gilles Pinson's chapter embraces the considerably complex but likewise notably interesting French situation. Defending the perspective for a new form to look at urban governance in France – not so much through regulatory and functionalist analyses of vertical or centre–periphery relationships and combinations between the different layers of governmental power, followed by a somewhat secondary view of horizontal local governance stakeholdings – he writes upon the recent consolidation of French cities as much more empowered political actors. Concomitantly expressing that today it is the local level and the local actors which are fully responsible for the ways urban policy is oriented and implemented in each city, placing much significance on local governance interactions, on specific local actors and socio-cultural structures, on political economy local strategies and influences, and thus on corresponding forms of urban regimes. Pinson concludes therefore that the regime approach shows itself to be one of the most interesting forms to deepen the comprehension of the political tendencies of such particular urban societies as those of the varied Southern European cities.

Ioannis Chorianopoulos begins his chapter with a reference to the fact that over the course of the twentieth century, and unlike what has happened in many other European countries, the significant growth of Greek cities is not owed to industrialisation processes. When Greece was granted European funds for urban development (namely with the URBAN programme) during the 1980s and 1990s, these were essentially structured and intended with a Northern European logic in mind, according to which the main issues to be addressed were the consequences of deindustrialisation on urban centres and suburbs. All in all, Chorianopoulos focuses on showing the impact of the regulatory and cooperative formulas imposed by the EU regarding the allocation and management of these funds. The strong interventionism of the central Greek state (in the decisions, control and implementation of funding), along with bureaucratic rigidity and no tradition of citizen participation and partnership with local and private initiatives, are constant features which also reflected the relatively reduced urbanistic transformations and the limited innovations ultimately applied to city management and local governance itself. In the support of these reflections, the author presents two case-studies addressing the cities of Keratsini-Drapetsona and Heraklion, both of which were granted EU funding at separate times – thus confirming not only some concrete positive progress verified on known socio-political inertial dimensions (both at local and national perspectives), but also the slow (yet positive) pace of transformation in Greek cities' local governance.

In her chapter, Irena Bačlija asserts that in the ever-changing environment of the city, urban governments are facing similar pressures as entrepreneurs in a competitive market. Following her view, the socio-political landscapes in which urban governments function is changing dramatically and the challenges that local officials meet shift accordingly. Resumed, it is the kind of shift that David Harvey has described as the pasage from managerialism to entrepreneurialism. For Bačlija it is indisputable that there is great demand for specialised tools which could help urban leaders and stakeholders to juggle multiple urban challenges while working towards long-term strategies and policies. Her chapter aims at presenting these tools through the lenses of urban governance and urban management, at the same time distinguishing urban management from urban governance. An empirical study on urban management in Slovenia (as part of a wider research on the whole EU) is presented as an fine example of the problems for the contemporary competitive city facing the challenges of administration reform.

Nil Uzun shows how in Istanbul the restructuring and reinforcement of urban governance positioning and corresponding dynamics are evident. Although not directly influenced by elements that are as relevant to its Southern European sister cities, such as EU policies and processes, the large metropolis of the Bosphorus is certainly influenced by globalisation issues, such as those concerning political economy, state devolution and civic empowerment. Here the repositioning of the city as a stronger political actor seems to have also taken place, along with greater local socio-political and cultural complexity. Nonetheless these tendencies do not show clear-cut conclusions, namely if the socio-political milieux in Istanbul are heading towards more effective and democratic policy deliveries. And especially when partisan and patronage stakeholding might well be reinforcing positions. Uzun's reflections upon Istanbul and its socio-political developments lead us to consider even more the perspective that the debates and questionings around the theme of urban governance and urban regimes are highly important both for deeper analytical research and for policy criticism.

The volume is completed with a more general concluding text written by the editors. The final chapter re-opens and somewhat synthesises most of the main tendencies facing Southern European cities' governance. As understood, the enlightening perspectives of cooperation, participation and collective construction have been increasingly accompanied by shadowed fears of public demission, oligarchic regimes and less local democracy. These lights and shadows of urban governance and the dilemmas they bring are particularly relevant to the cities of Southern Europe, whose socio-cultural specificities (namely considering their complexity and organic characters) very much determine local political and policy materialisations. The EU has been an important factor in bringing objectivity and rationality to public policy and governance networking – notwithstanding remaining a considerably distant frameworker. The text conjoins the relevance of specific urban Mediterranean socio-political and cultural perspectives – including when gaining cosmopolitanism and thus in a certain sense reducing North–South dualisms – and thus proposes a systematisation of both governance tendencies and concomitant areas for deeper analysis and reflection upon the Euro-Mediterranean urban world.

An Urban and Road Order

We remember well our fascination when we heard from a local public officer that in a single kindergarten in the centre of Barcelona the children spoke 25 different languages. Undoubtedly a completely new sort of cosmopolitanism is presently developing in the Mediterranean world, an evolution that will shape its socio-cultural landscapes in the near future, providing not only different lifestyles but also differential and much more demanding social movements and civic needs in the cities of Southern Europe.

The *spirit of place* of the Mediterranean world has long been under a socio-cultural complexification trend. As Matvejevitch (1992) wrote in his passionate book, the Mediterranean 'is not only a geography'. The profound cultural heritage and the corresponding complexity of the symbolic and socio-political reflections have for many centuries developed a landscape where 'the tendency to confound the representations of reality with reality itself tends to perpetuate ... an amplified identity of being that surpasses the identity of doing, not so well defined'. Here, governance is not at all a mere projection of objective project-driven and resource-capture strategies and stakeholdings – it is above all a reflection of social and cultural stratification.

The theme here proposed for reflection – urban governance in the Southern European cities – views the city especially as a place of power, of pilgrimage and of interchange. Most notably, as a place of political empowerment – thus comprehending the roads that configure its connections, its networks and its stakeholdings. The way Braudel expressed the human order of the Mediterranean world – *an urban and road order* – highlights the relevance of better understanding its urban governance panoramas and its respective lights and shadows.

Amidst the most marking traces of the Southern European cities might be their characteristics as trading hubs for goods, ideas and cultures – joining three vital connecting elements: a maritime (today globalised) expanse for commerce and interchange, rich hinterlands, and a large and varied population available for relational and trading activity – thus continuously forming dynamic spaces of interchange, as much of passage as of permanence. Formerly a strong civilisational root, today the Mediterranean world and its crossroads are positioned between different cultures, societies and ways of development, positioned as a trigger territory for humanity. The choices might be between a dispersed, individualist and fearful future, or a cosmopolitan, diverse and connected world. In a most crucial manner, the stages where these choices will be made are undoubtedly the stages of its cities – once again civilisational hubs.

References

Bagnasco, A. and Le Galès, P. eds. 2000. *Cities in Contemporary Europe.* Cambridge: Cambridge University Press.

Bourdieu, P. 1997. The forms of capital, in *Education: Culture, Economy, Society*, edited by A. Halsey, H. Lauder, P. Brown and A. Wells. Oxford: Oxford University Press, 46–58.

Braudel, F. 1949. *La Méditerranée à l'époque de Philippe II*. Paris: Armand Colin.

Brawley, L. 2009. The practice of spatial justice in crisis. *Justice Spatiale/Spatial Justice*, 1 [www.jssj.org].

Brenner, N. 2004. Urban governance and the production of new state spaces in Western Europe, 1960–2000. *Review of International Political Economy*, 11, 447–88.

Clark, T. and Hoffman-Martinot, V. 1998. *The New Political Culture*. Boulder: Westview Press.

Coleman, J. 1990. *Foundations of Social Theory*. Cambridge: Belknap.

Coletta, T. 2008. Une Reflexion sur l'Esprit du Lieu de la Ville Méditerranéenne. *16th General Assembly and Scientific Symposium*. Québec (Canada).

Hall, S. and Massey, D. 2010. Interpreting the crisis. *Soundings*, 44, 57–71.

Jessop, B. 1998. The rise of governance and the risks of failure: the case of economic development. *International Social Science Journal*, 155, 29–45.

Jessop, B. 2002. Liberalism, neoliberalism and urban governance. *Antipode*, 34, 452–72.

Jouve, B. 2003. *La Gouvernance Urbaine en Questions*. Paris: Elsevier.

Leontidou, L. 1990. *The Mediterranean City in Transition — Social Change and Urban Development*. Cambridge: Cambridge University Press.

Leontidou, L. 2010. Urban social movements in 'weak' civil societies: the right to the city and cosmopolitan activism in Southern Europe. *Urban Studies*, 47(6), 1179–203.

Maloney, W., Smith, G. and Stoker, G. 2000. Social capital and urban governance: adding a more contextualized top-down perspective. *Political Studies*, 48, 823–84.

Massey, D. 2011. Ideology and economics in the present moment. *Soundings*, 48, 29–39.

Matvejevitch, P. 1992. *Bréviaire Méditerranéen*. Paris: Fayard.

Mayer, M. 2007. Contesting the neoliberalization of urban governance, in *Contesting Neoliberalism: Urban Frontiers*, edited by H. Leitner, J. Peck and E. Sheppard. New York: Guilford Press, 90–115.

Mozzicafreddo, J., Gomes, S. and Baptista, J. eds. 2003. *Ética e Administração — Como Modernizar os Serviços Públicos*. Oeiras: Celta Editora.

Pace, G. 2002. Ways of thinking and looking at the Mediterranean city. *MPRA Paper*. Munich: University Library of Munich.

Putnam, R. 1993. *Making Democracy Work: Civic Traditions in Modern Italy*. Princeton: Princeton University Press.

Chapter 1
The Improbable Metropolis: Decentralisation and Local Democracy Against Metropolitan Areas in the Western World[1]

Christian Lefèvre[2]

Introduction

In the most recent literature on economic geography, urban planning, urban sociology or political science, metropolitan areas or city-regions are presented as the new 'spatial fix' (Harvey 1985) of the present period of capitalism that is globalisation. In other words, metropolisation is viewed as a process very much connected with globalisation in which city-regions are put forward as loci where the most salient societal issues are taking place: economic growth and wealth, social inequalities, environmental degradation, multicultural integration and so on.

Saskia Sassen in her seminal work (1991) showed that some cities concentrated headquarters and executive offices of some crucial international activities, notably in the finance, insurance and real estate sectors, and as such were becoming places of command in the global economy. She thus identified three 'global cities', London, New York and Tokyo. Although, or because, Sassen's theory has very much been criticised, it paved the way for a long series of research in the fields of urban economy and geography. All have subsequently demonstrated the importance of the largest metropolitan areas for the economic development of the world for various reasons. Allen Scott (1998) has shown that city-regions were attractive for firms because they provided low transaction costs and this largely explained their concentration in metropolitan areas. Michael Storper (1997) has stressed the significance of non-market interdependencies to explain why city-regions were so appealing to business. Pierre Veltz (1996) has presented city-regions as places offering what he called *assurance-flexibilité* [insurance-flexibility] for enterprises but also for individuals, meaning by this expression that metropolitan areas are attractive because they provide firms and people with choices, alternatives and opportunities (in finding jobs, in finding

1 An early version of this text was published in 2010 in *Análise Social. Revista do Instituto de Ciências Sociais da Universidade de Lisboa*, 197, 623–37.

2 IInstitut Français d'Urbanisme; Université Paris-Est Marne la Vallée; 5 boulevard Descartes, Champs sur Marne; 77454 Marne la Vallée Cedex 2 (France). christian.lefevre@univ-mlv.fr.

the appropriate qualified staff, etc.) that no other territories could offer. All these works were largely corroborated by sophisticated comparative international data, studies and rankings produced by Peter Taylor's team in his Globalization and World Cities Research Network (Taylor 2003).

In all these works, the political dimension of the metropolisation process as a new spatial fix of global capitalism is absent either because it is not taken into consideration or because it is viewed as automatic. For instance, Allen Scott (1998) assumes that once a city-region[3] develops economic agency, political organisation will automatically follow in a rather functionalist way.

If city-regions are relevant and crucial spaces for the production of actions and policies necessary to deal with most important societal issues (Rodríguez-Pose 2008), this means they must be governed for these policies to be produced and implemented. To be governed, they must become political spaces.

What is a political space? We can define it as a space of involvement of political, economic and social players (Cox 1998) where a legitimate collective action is produced, an action necessary to address existing issues and orient the future. Following Boudreau and Keil (2004), a political space contains three inter-related elements: a) a political and institutional entity; b) public policies; c) modes of social regulation. Regarding city-regions, by political and institutional entity we mean any political and institutional structure or arrangement at the metropolitan scale possessing political legitimacy and responsibilities; by public policies we mean the production of policies dealing with societal challenges and problems and their implementation at the metropolitan scale by various actors (states, local governments or any other public bodies); by modes of regulation we mean the existence of structures, arrangements, mechanisms and instruments at the metropolitan level capable of producing the mobilisation of actors, creating mediation between actors, allowing processes conducting to the production of collective action at the metropolitan scale.

The question of city-regions as political spaces is not new. Already in the 1960s, the 'Reformers' (Wood 1958), considering metropolitan areas emerging as social and economic spaces notably because of the evolution of transport and communication technologies, forecast that such an economic and social 'community' should have a political representation. But they assumed this political representation could not be the 'natural' result of the evolution of societies and cities and as such should be imposed. The history of metropolitan reforms in the United States and in Europe, which is largely a history of failures, proved they were wrong (Lefèvre 1998). However, they were not wrong in their diagnosis (the making of a new political space is not automatic) but in the way they wanted to create it (i.e. a top-down imposition) because the making of a new political space is inherently a conflicting process.

3 In this chapter, city-regions and metropolitan areas refer to the same space and scale. For a presentation and a discussion of the various expressions used to define these spaces and scales, see Rodriguez-Pose (2008).

This is indeed the focus point of both the work of Boudreau and Keil and of our own. Boudreau and Keil apprehend the production of new political spaces as a conflicting process. For them, 'new political spaces are the result of power struggles for constituting coherence and common objectives' because they challenge already existing political spaces (the state, the municipality, etc.). As such, the making of city-regions as new political spaces is the result of conflicts between actors and interests and by no means the logical result of the economic agency that city-regions have gained from the process of globalisation.

In this chapter, we will carry this idea of political spaces as a conflicting process in the case of metropolitan areas further by arguing, based on some empirical evidence from Southern European cities, that not only the Western experience has not attained success in this field but several present trends work in other directions. In the first section, we focus on the relationship between decentralisation and the building of metropolitan institutions showing that metropolitan areas have not been either the focus or the target of decentralisation processes and consequently have not gained much from this process. In the second section, we move onto the question of local democracy and show that this process has not favoured metropolitan areas either.

Decentralisation and Metropolitan Areas

The decentralisation processes that can be observed in most European countries and elsewhere in the world have not favoured metropolitan areas on the one hand because they have favoured other territorial scales (regions, provinces, municipalities) and on the other hand because the building of strong metropolitan authorities has been impeded by state and local actors.

Decentralisation against Metropolitan Areas

In most European countries, metropolitan areas have been the 'forgotten territories' of decentralisation. Generally speaking, decentralisation laws and decentralisation processes have transferred responsibilities and resources to already existing governmental tiers, that is, municipalities and provinces and in some countries to regions as well. Although the 'metropolitan fact' has emerged as a strong socio-economic and spatial phenomenon, it has not had any significant political and institutional responses as we shall see. In the UK, decentralisation – understood as a devolution process – has been given to 'peripheral regions' such as Wales, Scotland or Northern Ireland. In England, the most important attempt to decentralise at the regional level was killed off in 2004 when voters in the north-east strongly rejected a referendum to create a directly elected regional council. Since then, the process has stagnated apart from the relative exception of London. But the establishment of the Greater London Authority (GLA) in 1999 must not be seen as a sign of decentralisation towards the metropolitan level mostly because in the British institutional system, Greater London is indeed a region and the London situation was seen as a pioneering step towards a more general political regionalisation which has not been pursued so far.

Elsewhere the situation is approximately the same. In Germany, the Netherlands, Belgium or the Scandinavian countries, decentralisation, albeit timid in some cases, has favoured the regional level (Germany and Belgium) or counties (Scandinavia and the Netherlands).

The experience of Southern Europe (France, Spain and Italy) confirms this. In France, since the first decentralisation laws of the early 1980s, the state has transferred responsibilities and resources to all local governmental tiers: regions (created in 1982), *départements* and municipalities. As a whole the various decentralisation laws have been very careful to distribute more or less evenly the various transfers of powers among local governments. In this process, metropolitan areas have been 'forgotten' until very recently (2009) but so far no significant changes have yet been made. Generally speaking, the 'metropolitan phenomenon' has been institutionally – and in rare instances in public policy-making as well – addressed through *intercommunalités*, i.e. the voluntary grouping of municipalities belonging to a same urban area. The last *intercommunal* Act, passed in 1999, established new *communautés urbaines* for areas grouping more than 500,000 inhabitants (hardly a metropolis by international standards) but these structures are closely politically controlled by municipalities.

The innovation may come from the last proposal made in May 2009 by the Commission Balladur on Territorial Reforms which proposed establishing 11 *métropoles* in the 11 largest urban areas, with these *métropoles* being local government authorities in their own right with their own directly elected councils and significant responsibilities and fiscal and financial resources. It remains to be seen whether this proposal will be implemented or will be lost in political debate.

There is one major exception to this, the Paris-Ile-de-France region, which is by far the only French metropolis of international ranking. In this territory, decentralisation has always been less important than in the rest of the country and the state has retained major responsibilities and control over the development of the area. In most recent years, although several laws have transferred new powers (planning, public transport) to the regional level, the trend seems towards a 'return of the state' with reforms pushing towards a re-centralisation, one good example of this being the establishment of a 'ministry for the capital region' in 2008.

In Spain, the decentralisation process has strongly benefited regions, the 'autonomous communities', to the extent that Spain is today a quasi-federal country. However, the downwards pursuit of decentralisation has not benefited the metropolitan areas, on the contrary. To start with, one of the first actions taken by the Spanish regions was to abolish the existing metropolitan authorities, established during the Franco period. Thus, the Basque region abolished the *Corporación Metropolitana de Bilbao* in 1980, the Valencia region got rid of the *Corporación de la Gran Valencia* in 1986 and one year later the Generalitat de Catalunya eliminated the Metropolitan Authority of Barcelona. None of those metropolitan structures were replaced by democratically elected institutions of the same dimension. Second, the next step in Spanish decentralisation that is pursuing decentralisation processes below the regional level, the so-called *pacto local* has not taken metropolitan areas into consideration. Although with great difficulties and conflicts, this process has benefited municipalities and not the urban area as a whole. Finally, the most recent laws and national reflections dealing with cities (the 2003 Act for large cities and the 2005 local government white paper) hardly consider the metropolitan scale, except for the white paper suggestion of establishing 'metropolitan agreements' on a voluntary basis. In fact, the 2003 Act was more interested in strengthening the powers of central cities than addressing the metropolitan issue.

In Italy, decentralisation has been following a very long and winding path but has benefited all traditional local government tiers, from the regions, established in the 1970s, to the provinces and the municipalities. The process has been and still is rather confusing but, once again, metropolitan areas have not been favoured. On the one hand, it is true that the Italian constitution introduced 'metropolitan cities', i.e. metropolitan authorities, as part of the Italian Republic, thus giving metropolitan areas a constitutional legitimacy. But, on the other hand, these 'metropolitan cities' do not exist. Indeed, their establishment has been on and off the political agenda for about two decades now (since the 1990 142 Act) but none has been formed as we shall see in the next section. In the late 1990s, Italy was heading towards regional level federalism and part of the political elite seriously envisaged the formation of a national senate body composed of only regions and metropolitan cities. This would have given the metropolitan areas strong political recognition but did not happen for several reasons, among them the political turmoil of this period and the opposition of traditional local governments like the municipalities and in some cases provinces. As a result, those institutions which benefited from decentralisation laws (such as the Bassanini laws of the late 1990s) were those already existing, regions, provinces, municipalities and not the metropolitan areas.

The Failure of Building Metropolitan Authorities

In Europe – although the situation is similar elsewhere in the world – attempts to build metropolitan authorities, that is, local government units covering more or less the urban area and benefiting from political legitimacy with significant and adequate responsibilities and resources (Sharpe 1995), have been many. By and large, they have not met any real success (Lefèvre 1998, 2008, 2009) and in most 'successful' cases, these authorities have been weak. At least three major reasons explain this. First, states have been unwilling to decentralise at that level because they have been and remain afraid about establishing strong political counter-powers to their authority. This is all the more the case when dealing with metropolitan areas which are at the same time the capitals of their respective countries (Lisbon, London, Paris). Second, generally speaking, local governments belonging to the metropolis have opposed the establishment of such authorities also out of fear of losing powers and having actions and policies imposed by those metropolitan bodies. Third, when these authorities have been established, they have encountered the rivalry of central cities which have been able to significantly reduce their juridical powers. We illustrate this in the following section by focusing on Southern European countries (France, Italy, Portugal and Spain).

France may be described as the 'good pupil' of metropolitan government because – with the important exception of the Ile-de-France area – all major big cities possess their own metropolitan authorities: the *communautés urbaines* for the largest and the *communautés d'agglomération* for those with between 50,000 and 500,000 inhabitants. In this chapter, we focus on the largest urban areas.

The eight largest cities (Lille, Lyon, Marseilles, Nice, Strasbourg, Bordeaux, Nantes, Toulouse) are all covered by a *communauté urbaine*. Such a body is a grouping of municipalities (a grouping imposed by the state in Bordeaux, Lille, Lyon and Strasbourg at the end of the 1960s, on a voluntary basis for the others) which by law has responsibilities for most policy sectors of metropolitan interest (public transport, economic development, planning, waste management, etc.) and financial and fiscal resources of its own to carry out these responsibilities. In theory, *communautés urbaines* can be considered strong metropolitan authorities. When looked at closely, the situation is different.

First in terms of their geographical scale, most *communautés urbaines* do not cover their real functional areas (measured by daily trip patterns for instance), the reason being that most of them were established at the end of the 1960s and have not expanded their territorial range since, although urbanisation was already taking place in that period. Second, in political terms, municipalities belonging to the same metropolitan area, whatever their political partisanship, have agreed to limit the powers of those authorities and have been able to do so because they control the boards of the *communautés*, very often dominated by the central city. The rule has been that the *communautés* should not impose any decision or policies on a single municipality. As a result, until very recently, the *communautés* have been politically very weak and have not been able to produce and implement metropolitan policies in most cases. This situation has been constantly denounced by several national reports and reviews (Dallier 2006), accusing municipalities of getting together more to benefit from central government financial help[4] than to work collectively. One of the most illustrative examples of such a failure is the Marseilles metropolitan area situation, where, although this area is functionally completely integrated, it is 'administered' by no less than four *communautés*, each controlled by a central city (the most important being Marseilles itself) and each pursuing its own strategy and its own policies.

In Spain, as we have seen, metropolitan corporations were abolished in the 1980s and have not been replaced by metropolitan authorities since. True, the

4 When municipalities form a joint authority with its own fiscal resources (which is the case of *communautés urbaines*), the state gives this joint authority a significant financial bonus.

Corporación de la Gran Valencia was replaced in 1986 by the *Consell Metropolità de l'Horta*, a much less powerful body, but this council was also abolished in 1999. In Barcelona, the *Corporació Metropolitana de Barcelona* was replaced in 1988 by a *Mancomunidad*, i.e. a joint authority grouping 31 municipalities essentially in the domains of urban planning and land protection. This body is very weak and is chaired by the city of Barcelona like other smaller structures such as the Metropolitan Transport Entity. By and large, the metropolitan area of Barcelona has no metropolitan authority (however see the next section).

The same can be said of all the largest Spanish urban areas with the exception of Madrid. Indeed, Madrid may be the only big world city with a metropolitan government in its own right. However, this metropolitan authority – the Autonomous Community of Madrid, the CAM – was established by chance. The creation of the CAM in the early 1980s was the result of a political compromise between the young political parties of that period since Madrid was neither a natural nor a historical region of Spain. The compromise was to set up a new region which would cover the municipality of Madrid and what was left from the establishment of the surrounding regions (Castile-La Mancha, Castile y León). In that compromise, the idea of giving Madrid an institution covering the functional area never came about. It is thus by chance that the Madrid metropolitan area got a regional body which in the long run proved large enough to envelop the growing metropolitan area, and since in the 'State of Autonomies', Spanish regions are strong institutions, almost federated states in federal countries, the Madrid area got a strong metropolitan authority (Rodríguez-Álvarez 2002).

In Italy, the issue of metropolitan government was directly tackled by Act 142 in 1990. In this Act, *città metropolitane* (CM) were envisaged for the 10 largest urban areas. CM would be new local government units, covering the whole urban area and possessing area-wide competences (public transport, planning, urban development, physical networks, etc.). They would be administered by directly elected councils. Act 142 envisaged the merger of small municipalities within the respective metropolitan areas as well as the splitting up of central cities and the substitution of provinces covering these major urban areas with the new *città metropolitane*.

Considering the direct attack against local governments and notably central cities and in some instances provinces, it is no wonder that Act 142 was never implemented. Several attempts were made, the most innovative in Bologna (Jouve and Lefèvre 1996), but none succeeded and Italian cities still do not possess *città metropolitane*, after about two decades of Act 142. The failure of Act 142, as far as the building of metropolitan authority is concerned, is largely due to the opposition of central cities and provinces which strongly resisted the implementation of CM on the very grounds that this implementation would mean their total disappearance. For instance, Milan was to be split up into 10 new municipalities and Bologna into seven. Although this process was supported for a while by the municipality of Bologna but for very specific reasons (Jouve and Lefèvre 1996), it was strongly rejected by the city of Milan. In other metropolitan areas, the idea of building CM did not even go through a debate stage.

Special mention must be made of the capital, Rome. Although concerned with 142 Act, no CM was established either for that territory in spite of various proposals made by the central city, notably during Veltroni's mandates. However, in 2001, the Italian government inscribed the issue of metropolitan governance of the capital into the constitution. Since then, nothing has happened due to the opposition of the regional and provincial councils. However, things began moving again in late 2008 with the creation of various local and national working groups and commissions in charge of debating and implementing the *Città metropolitana per Roma capitale* and the declared willingness of central government to go ahead with such an initiative. It remains to be seen whether these moves will be sufficient to establish such a metropolitan authority.

Finally, in Portugal, although the urban areas of Lisbon and Porto have had metropolitan governments since the early 1990s, these are relatively weak. They are weak because their political legitimacy largely lies on the member-municipalities which control both the metropolitan authority's council and its executive (e.g. the *junta* is composed of the presidents of the member-municipalities). In that context, central cities have resources to oppose metropolitan interests as is the case in the capital (Nunes Silva and Syrett 2006). They are also weak because the state remains in charge of several significant domains among which are transport, economic development and planning.

Building Metropolitan Government by Governance: Uncertain 'Success'

Faced by the abolishment of metropolitan authorities and/or the incapacity of political systems to establish any such structure, some urban areas (Barcelona, Bilbao, Bologna, Florence, Turin, Venice among others) have launched experiences based on 'procedural policies' (Duran and Thoenig 1996). These experiences have aimed at the establishment of metropolitan forms of government through a complex process of coalition building and project elaboration through specific instruments and arrangements. Strategic planning has very often been the policy domain used to monitor these processes. In Europe, the two emblematic cases of such experiences are Barcelona and Turin but it remains to be seen whether these experiences have been successful because for the time being both Barcelona and Turin are still waiting for a metropolitan authority to emerge.

Barcelona is very famous for its pioneering role as a 'strategic city'. The first strategic plan was indeed launched in 1988 in preparation for the 1992 Olympic Games. It was followed by two other plans which had only a municipal dimension. In 2000, the first metropolitan strategic plan was approved; it concerns the metropolitan area of Barcelona (AMB), more than three million people spread out over 31 municipalities. The metropolitan strategic plan is not a master plan but more an 'orientation' plan which has set guidelines for the development of the whole area. It is 'managed' by a complex structure made up of a General Assembly composed of all the major actors of the metropolis (about 300 members including the chamber of commerce, universities, the city of Barcelona and the

30 other municipalities, the joint authorities, banks, business associations, unions, cultural associations and foundations), an executive commission of 30 people, representing the most important stakeholders, in charge of plan administration. This commission is assisted by several committees and working groups whose missions are: a) to feed the plan with reflections and data and b) to ensure that the plan is linked with sectoral plans (transport, environment, land use, housing, etc.) and existing procedures of cooperation. This complex arrangement is chaired by the Mayor of Barcelona and is used as a tool to mobilise the whole metropolitan society. It is through this mobilisation and this 'organisational engineering' that political actors can 'govern' the metropolitan area.

In many ways, the situation is similar in Turin. Here also, there has been a first strategic plan approved in 2000 and a second one in 2006. These two plans cover the whole metropolitan area and are managed by a structure comparable with that of Barcelona: the 'metropolitan assembly' is composed of 122 members (the suburban municipalities, the central city, the Province, business associations, chamber of commerce, unions, cultural foundations, etc.). An executive committee of 10 persons ensures day-to-day decision-making and a specific agency, *Torino Internazionale*, is in charge of day-to-day plan administration. Both the assembly and the committee are co-chaired by the Mayor of Turin and the President of the Province. However, one single feature differs significantly from the Barcelona experience. In the case of Turin, the building of a metropolitan government as an institution has been one of the aims of strategic planning. To this purpose, a metropolitan conference composed of the 39 municipalities and the Province was established in 2000. The idea was to use the strategic planning process as a feeder for a metropolitan development project, this process resulting in the establishment of a metropolitan authority, of the 'citta metropolitana' type, for instance. So far, the process is still on but no significant moves towards the building of such an authority are to be seen and the creation in 2007 of a 'metropolitan table' composed of 17 municipalities may only be interpreted as a setback.

Local Democracy and Metropolitan Areas

Local democracy considered as the development of institutional arrangements to enhance the involvement and participation of citizens and civil society in local affairs has significantly expanded in recent decades in Europe. Presented almost everywhere as a sign of a more politically vibrant society and an instrument which allows us to address existing problems more successfully, local democracy can nevertheless be questioned in its relationship with the process of metropolisation. In other terms, the modalities of local democracy development may be analysed as being at odds with the making of metropolitan areas as political actors.

The Development of Local Democracy in Europe

Local democracy usually is pursued through two types of instruments: a) the making of infra-municipal institutions and b) the direct involvement of citizens through the procedures of referendum and initiatives and the establishment of neighbourhood councils. Both instruments have been used and developed in European countries.

In many metropolitan areas, infra-municipal institutions have been created in the last decades. Although they may vary in their responsibilities, political legitimacy and resources, they have developed in almost every country. For instance, one finds 12 *bezirke* in Berlin, 15 *stadsleden* in Amsterdam, 21 *distritos* in Madrid, 19 *municipi* in Roma or 20 *arrondissements* in Paris. These institutions are generally administered by locally elected councils and receive their budget from their municipality. In more recent years, they have spread over many more cities but one common element is the fact that they are usually limited to the central cities of the largest urban areas.

In addition to these institutions at the infra-municipal level, many countries have set up other structures aiming at the direct involvement of inhabitants but usually on a smaller scale, a type of neighbourhood council. One finds this type of council in many Italian cities, in some London boroughs like Tower Hamlets or Islington, in Copenhagen or more recently in France in all municipalities with over 80,000 inhabitants, as this is mandatory by the 2002 Act on 'democracy of proximity'. Generally, these bodies are politically weak, partially appointed by the municipality, and have no decision-making capacity because they are essentially consultative.

Finally, the procedures of local referenda and initiatives, once restricted to a few countries like Switzerland in Europe or the United States, have proliferated in Europe in recent years. In Germany, all Länder have now introduced these measures into their constitutions. In Italy and the Netherlands, these instruments are ever more commonly used at the local level. Even in countries which have traditionally been the bastions of representative democracy and which have opposed these procedures, such as France, local referenda and initiatives have been made not only legal but have been given a decision-making character in some specific cases. In France, local referenda were legalised in 1992 and the 2003 and 2004 Acts have made them decisional (hence, when the referendum is approved, it becomes law).

If the development of local democracy instruments can be regarded as theoretically positive because it enhances the involvement of citizens and thus contributes to making municipalities more democratic, the question remains as to the political existence of the metropolitan scale and territory in that context.

Local Democracy against the Metropolis?

The vast majority of local democracy instruments established in the recent period have not focused on the metropolitan level. By and large they have privileged smaller scales to the extent that it may be argued that they have politically strengthened non-metropolitan territories and in some instances have been used against the political recognition of the metropolitan area.

First, one major result of the development of these forms of local democracy has been the strengthening of infra-municipal territories. Certainly, this is at the same time logical and inevitable since these local democracy instruments have been created for that purpose. However, the impact of the strengthening of infra-municipal territories on the metropolitan level is ambiguous and must be clarified.

On the one hand, it is possible to argue that the strengthening of infra-municipal territories is an obstacle to the emergence of the metropolis as a political actor because these infra-municipal territories will tend to use their new powers and resources to get more autonomy from their municipality and above, from the metropolitan area, and at the same time because they contribute to increasing political and institutional fragmentation. The Berlin and Rome cases are good examples of such a situation because *bezirke* and *municipi* have constantly gained powers and resources over the years to the extent that they have been able to challenge the power of the central city and in the case of Berlin to isolate themselves from it (Röber and Shröter 2007). In Rome, it is expected that the *municipi* will become municipalities in their own right in the framework of a possible *città metropolitana*.

Of course, the political development of infra-municipal territories is not automatically synonymous with autonomisation or the separation from the central municipality as long as metropolitan counter-forces and instruments are established in order to preserve the whole. But these counter-forces or instruments are not to be seen in the various institutional reforms and therefore the risk is high of further fragmentation and autonomy due to the use of powers and resources given by decentralisation and local democracy instruments to infra-metropolitan bodies.

One good example of such a risk is given by the many referenda which have been used to secede from the central city or to oppose the establishment of metropolitan authorities.

The country in which local referenda have been most widely used on the issue of metropolitan government has been the United States. For several decades, citizens of many metropolitan areas have been asked to approve or to oppose the establishment of metropolitan authorities. In general, they have opposed such a creation, which partially explains the very small number of US metropolitan areas which possess some sort of metropolitan government.

But the United States is not the only country where such a situation has occurred. In Europe as well, local referenda have been used to reject metropolitan governments. One can mention the largely negative referenda on the establishment of 'city-provinces' in Amsterdam and Rotterdam in 1995. One can also cite the negative referendum regarding the merger of the Länder of Brandenburg and Berlin in 1996 which would have allowed the setting up of a de facto metropolitan Land over the Berlin area.

To be fair, the successful referendum over the creation of the Greater London Authority (GLA) in 1999 must also be referred to but this positive result needs to be judged with caution because the London situation is very peculiar. First, because it is the only metropolis which has had a metropolitan authority for a long time (the London County Council was created in 1889, it was then followed by the Greater London Council, abolished in 1986). Thus, the establishment of the GLA was only somewhat a 'return to the past', to a situation that Londoners had known for quite a while. Second, because London has no central city and the conflicting relationships between the centre and the periphery have less institutional grip there than in other metropolitan areas.

Local referenda have also been used to secede from existing municipalities, usually central cities, which is both constitutionally possible in the United States and generally accepted by state legislatures. For example, in the 1980s, West Hollywood seceded from Los Angeles. In 1993, the voters of Staten Island, one of the five boroughs of New York, approved their secession from the city although this secession was later rejected by the state. Several attempts have been tried in Los Angeles in the decade after 2000.

This phenomenon of secession from municipalities or even metropolitan authorities has been experienced in other countries as well, with relative success. One may mention the successful referendum in Montreal in 2003 which allowed the newly merged municipalities to the Metropolitan city of Montreal to 'de-merge', which some of them have done. One may also point out the various unsuccessful referenda asking the voters of Mestre to secede from the municipality of Venice (Italy) in recent years.

All these examples are pointed out not to conclude that the development of local democracy, per se, runs counter to the political emergence of metropolitan areas, but to stress the risk that this development may incur as long as local democracy measures and instruments do not take the metropolitan scale into consideration, measures and instruments which would act as counter-forces to the elements of 'nimbysm' or localism that local democracy inevitably bears. It is also arguable that such situations have not been frequent because in Europe, contrary to the United States, national constitutions usually do not grant such 'secessionist' powers to citizens.

Conclusion

The making of metropolitan areas as political spaces is a conflicting process as we have seen. In this chapter, we have focused on two major obstacles preventing such a process from succeeding: the opposition of political-institutional actors such as the state and local governments in their use of decentralisation and the ways local democracy is developing and practised, at least in Europe and North America. However, coming back to the inter-related elements necessary for a political space to exist according to Boudreau and Keil (2004), the importance of other actors such as business and civil society should be mentioned although the experiences of Barcelona and Turin presented above may be interpreted with caution.

References

Boudreau, J.A. and Keil, R. 2004. In search of a new political space? City-regional institution-building and social activism in Toronto. *Annual Meeting of the Association of American Geographers*, Philadelphia, United States, 15–19 March.

Cox, K. 1998. Spaces of dependence, spaces of engagement and the politics of scale. *Political Geography*, 17(1), 1–23.

Dallier, P. 2006. *Rapport d'Information sur l'Intercommunalité à Fiscalité Propre*, 193. Paris: Sénat.

Duran, P. and Thoenig, J.C. 1996. L'Etat et la gestion publique territoriale. *Revue Française de Science Politique*, 46(4), 580–623.

Harvey, D. 1985. *The Urbanization of Capital*. Baltimore: The Johns Hopkins University Press.

Jouve, B. and Lefèvre, C. 1996. Dynamiques institutionnelles et culture politique territoriale: la cité métropolitaine de Bologne. *Revue Française de Sociologie*, 37(3), 369–96.

Lefèvre, C. 1998. Metropolitan government and government in Western countries: a critical review. *International Journal of Urban and Regional Research*, 21(4), 9–25.

Lefèvre, C. 2008. The democratic governance of metropolitan areas: international experiences and lessons for Latin American cities, in *Governing the Metropolis*, edited by E. Rojas, J.R. Cuadrado-Roura and J.M. Fernández-Güell. New York: David Rockefeller Center for Latin American Studies, Harvard University.

Lefèvre, C. 2009. *Gouverner les Métropoles*. Paris: LGDJ-Dexia.

Nunes Silva, C. and Syrett, S. 2006. Governing Lisbon: evolving forms of city governance. *International Journal of Urban and Regional Research*, 30(1), 98–119.

Röber, M. and Shröter, E. 2007. Governing the capital: comparing institutional reform in Berlin, London and Paris, in *Governing Cities in a Global Era: Competition, Innovation, and Democratic Reform*, edited by R. Hambleton and J. Gross. New York: Palgrave Macmillan.

Rodríguez-Álvarez, J.M. 2002. Madrid: le pilotage politique par la région, in *Métropoles Ingouvernables: les Villes Européennes entre Globalisation et Décentralisation*, edited by B. Jouve and C. Lefèvre. Paris: Elsevier.

Rodríguez-Pose, A. 2008. The rise of the 'city-region' concept and its development policy implications. *European Planning Studies*, 16(8), 1025-46.

Sassen, S. 1991. *The Global City. New York, London, Tokyo*. Princeton: Princeton University Press.

Scott, A. 1998. *Regions and the World Economy: The Coming Shape of Global Production, Competition, and Political Order*. Oxford: Oxford University Press.

Sharpe, L.J. 1995. *The Government of World Cities: The Future of the Metro Model*. Chichester: John Wiley & Sons.

Storper, M. 1997. *The Regional World: Territorial Development in a Global Economy*. New York: Guilford Press.

Taylor, P. 2003. *World City-Network*. London: Routledge.

Veltz, P. 1996. *Mondialisation, Villes et Territoires*. Paris: PUF.

Wood, R. 1958. The new metropolis: green belts, grass roots or gargantua. *American Political Science Review*, 52, 108–22.

Chapter 2

The Institutional Dimension to Urban Governance and Territorial Management in the Lisbon Metropolitan Area[1]

José Luís Crespo[2] and João Cabral[3]

Introduction

Governance is a long-standing term/concept and a still older reality (Peters 2002, Pierre and Peters 2000). Societies have always needed some form of orientation and guidance, leadership, and collective management. Variations in political and economic orders have produced different responses to fundamental issues relating to just how to guide and structure society and how best to meet the range of challenges resulting and generate the respective responses needed. In this sense, governance is no constant, as it tends to change to the extent that needs and values also change.

The usual responses to such questions were drafted by the state but whatever solutions may have proven effective within one particular context, soon turned out to be ineffective with the passage of time. The government/governance process represents a continuation of the joint set of policies of administrative adaptations and activities in reaction to changes in society, in terms of what is designed to represent a 'tailoring' of the means of development and the achievement of collective goals (Peters 2002).

The adaptative capacity of contemporary governance questions the assumptions upon which are based, and which regulate, those approaches deemed 'traditional', specifically as regards the centrality of state intervention and public authority in government. The notion of a single *locus* of sovereignty and a hierarchical structure to the governance system no longer corresponds to reality.

As a concept, governance emerges out of the shared conviction that, across various levels and degrees, 'the traditional structures of authority ... failed' (Kooiman 1993: 251) and that the modern state is now forced into a cycle of

---

1 An early version of this text was published in 2010 in *Análise Social. Revista do Instituto de Ciências Sociais da Universidade de Lisboa*, 197, 639–62.

2 CIAUD; Faculdade de Arquitectura; Universidade Técnica de Lisboa. Rua Sá Nogueira; Polo Universitário do Alto da Ajuda; 1349-055 Lisbon (Portugal). jcrespo@fa.utl.pt.

3 CIAUD; Faculdade de Arquitectura; Universidade Técnica de Lisboa. Rua Sá Nogueira; Polo Universitário do Alto da Ajuda; 1349-055 Lisbon (Portugal). jcabral@fa.utl.pt.

re-legitimation. The traditional conceptualisation of government, recognising the state as the most prominent actor at play in public politics, are considered as outdated approaches for the organisation of social interactions. These perspectives on governance, instead, seek to aggregate 'the totality of theoretical conceptions on governing' (Kooiman 2003: 4) and are considered an effective 'process of orientation for society' (Peters and Pierre 2003: 2).

However, there are no replacements or generally accepted alterations to guide and structure these new assumptions and, consequently, these have become even more problematic than before, for both the academic world and practitioners.

Against this backdrop, this chapter seeks to discuss the concept of governance across its various facets, in terms of both meanings and perspectives. Another interpretative component relates a range of phenomena and changes taking place in society with the concept of governance itself. We then move on to emphasise those governance perspectives that have most moulded the thinkings on urban policy. Finally, for the Lisbon Metropolitan Area, we analyse governance and territorial management, illustrating two critical aspects: a) the question of its role and its implications for public administration decisions at the central/regional and local levels in terms of territorial planning and development, and b) the emergence of municipal and inter-municipal institutions and companies with responsibilities for the management of areas and services within the scope of local administrative competences, with the goal of promoting flexibility, especially contractual and institutional interaction to generate greater profitability in providing these services.

Governance: The Adaptability of a Concept Containing Various Meanings and Perspectives

Governance is a very loose term, an 'umbrella' concept, sometimes badly interpreted due to the multiple meanings attributed to it (Pierre 2005). Indeed, one reason for its popularity is its ability – contrary to the more restrictive term governing/government – to cover the whole range of institutions and relations involving the governing process (Peters and Pierre 2000).

Initially, the concept was closely tied up with that of governing/government. In this limited sense, its utilisation for a long period was restricted to the juridical and constitutional field to describe the running of state affairs or the management of an institution characterised by a multiplicity of actors, where the expression 'government' seemed excessively restrictive. More recently, a majority of authors have related the concept with distinct analytical frameworks (Stoker 1998). It has mushroomed across the output and vocabulary of the social sciences as a term/concept in fashion in various fields: politics, economics and international relations, among others. The ideas associated are fairly diverse even if only referring to good governance, international, European, regional, metropolitan or urban governance, multi-level governance, vertical and horizontal governance – to mention but a few of the examples. Depending on interlocutors, their fields of activity or research or, more simply, their awareness, one or many of these possible meanings may be evoked (Borlini 2004).

This change in the meaning of government, associated with governance, falls within the framework of new governing processes. Governance refers to self-organisation, characterised by the interdependence of inter-organisational networks, recourse to exchanges in which the rules of the game endow some autonomy in relation to the state (Rhodes 1997).

Bevir et al. (2003: 45) defined governance as 'a change in the nature or meaning of government'. Correspondingly, recent years have seen the concept 'break free' of the shackles imposed by the aforementioned limitations to take on a broader meaning and field of application. This paradigmatic character enables its deployment across various fields. Kooiman (2003) identified no less than 12 different meanings[4] (depending on the respective field of usage), with only one set of common denominators – taking into consideration the main institutional spheres (the state, market and community). This broad range of meanings becomes apparent in the way the branches of the social sciences draw on different and, on occasion, distinct interpretations (Kjær 2004).

The modes of socio-political governance always result from interaction between the public and the private. Interactive socio-political governance involves defining the 'tone' and establishing the political-social conditions for the development of new interactive models that regulate through co-management, co-leadership and co-orientation (Kooiman 1993). Governance thus approximates the term coordination through the existence of coordinated actions dependent on contingent attitudes or institutional mechanisms for coordinating and concerting actors (Scharpf 2001).

According to Milward (2004: 239), governance is a broad-reaching term (and not only governmental) concerning the conditions for creating rules for collective actions, very often including private sector actors. The essence of governance is its focus on leadership and management mechanisms (subventions, contracts and agreements), which are not exclusively associated with either the entity or possible sanctions handed down by the government. These mechanisms or tools are deployed to connect the networks of actors operating across the different domains of public policies. One empirical issue is the extent of these operations, that is, just what autonomy do they enjoy, or are they state led. Governance means thinking about the means of orienting the economy and society, as well as the means of attaining specific collective goals in which the state plays a fundamental role, focusing on priorities and defining objectives (Peters and Pierre 2000).

More recently, another conception portrays how networking governance creates more opportunities for individuals to participate in political decision-

4 Minimum state, corporative governance, new public management, 'good governance', socio-cybernetic governance, self-organising networks, resource management, self-regulating societies, global governance, economic governance, governance and governability, European governance-multi-level governance, participative governance.

making processes and thereby become able to build social and political capital, as well as self-governance competences (Sorensen 2002, 2005, Sorensen and Torfing 2003).

In Europe, in recent years, the notion of governance, as a challenge to management and coordination, was the subject of widespread debate, both scholarly and political: a) governance taking a type of paradigmatic format, as a conceptual framework for posing a series of important questions about society but which, as this is posed in *pre-theoretical* terms, proves difficult to encapsulate (Le Galès 2003, Stoker 1998); b) other authors have already posited governance as *a theory* (Pierre 2000, 2005, Pierre and Peters 2000) – highlighting that while there is no complete theory of governance – this does constitute an analytical framework or, at the very least, a set of criteria that define those 'objects worthy of study' (Stoker 1998), within a perspective on governance that moves on from the assumption of institutions fully and exclusively controlling urban management and approaching them as a variable (Pierre 2005); c) in its *normative dimension*, governance is essentially bound up with a norm or an instrument of public utility, presented as some 'miraculous' solution, for example, in the analysis of public policies in accordance with those management trends seeking to raise the effectiveness and efficiency of public actions and that perceives governance as a tool for broadening participation in decision-making processes (Pierre 2005); and d) governance from an *analytical perspective* framed in support of an ideological discourse and as an insight for reading the transformations in public action, in particular public territorial action contributing to the new types of leadership/orientation that have been put into practice in recent decades across various domains or territories, bringing about the (de)institutionalisation of public policy analysis through focusing on a set of actors participating in the construction and treatment of collective problems (Leresche 2002).

Governance: Relationships between the Concept and the Field

We now proceed to set out some hypotheses on both the development and success of governance from the theoretical point of view and on the relationship between the concept and the set of phenomena it incorporates. In comparative empirical studies, the majority of interpretations are not sufficiently precise for differentiating between new forms of governance and traditional forms of government. Despite these divergences, there is a relative consensus around a certain number of core points.

There are countless studies on governance that effectively '*dress up mutton as lamb*' as many of the practices currently grouped under the governance concept were previously analysed from other perspectives, such as regulatory theories in terms of public and private partnerships, industrial districts, the study of networks, management and organisation, and the new public management (Borlini 2004, Jessop 1998, Le Galès 1995).

Governance and government are sometimes approached not as distinct entities but as *two poles on a continuum* of different types of government. While the extreme form of government was the 'strong state' in the era of 'big government' (Pierre and Peters 2000: 25), then the corresponding extremity in the form of governance essentially represents a self-organisation and coordination of the social actor network, in a sense of resisting government leadership and management (Rhodes 1997). The connotation of government and governance as the extremities of some theoretical *continuum* is, however, unlikely to prove sufficiently sensitive for capturing changes in the form and function of governance.

Part of the recent literature on governance stresses the importance of *multi-level governmental structures* to the dissemination of 'new' forms of governance. Its analytical focus is rather diffuse, concentrating primarily on the European Union (EU) level and without paying due attention to the way in which these 'new' forms of governance are (or might be) implemented at the member state and lower levels. Pierre and Peters (2000) hold that the state is losing its role as 'director' with control displaced to international and regional organisations, such as the EU, autonomous and municipal regions, to international corporations, non-governmental organisations, and other private or semi-private actors.

There is a close connection between the development and application of the governance concept and the social changes that have been taking place in recent decades with profound transformations impacting upon Western societies. This has led to the identification of a crisis in governability across various levels of government institutions, in effect, whether at the state or municipal level, *losing their capacity for action and for dealing with the ongoing transformations in society*. The processes of economic globalisation, European unification, the movement towards post-Fordist economic–social relations, demographic transformations, the added complexity of societies driven by their respective fragmentation, the unpredictability of the future, the lack of connectivity between weakened political authorities and citizens, and undermining – in terms of both its financial sustainability and legitimacy – the capitalist welfare system all go some way to explaining the failure of traditional models of public policies. The loss of this capacity to orientate led to a conviction that the state needs to alter the culture of its civil service (Pollitt 2000) or even delegate policies to actors beyond the state structure. Some authors conclude that both the opacity around the state (Rhodes 1997) and the formation of networks at various levels raise the complexity characterising modern society. All these facets render it increasingly difficult for the national state to enact its role as the unique regulator of the economy and social services, and increasingly demand that means of coordination and cooperation are established between institutions, territorial levels and differing actors (Le Galès 2003).

There is a tendency in the political science literature to associate government with regulation, while governance is frequently seen as a demonstration of the *appearance of new political instruments* (Jordan et al. 2005). Governance is thus characterised by the rising utilisation of non-regulatory political instruments. These are proposed, designed and executed by non-state participants, working either in conjunction with state actors or independently. We find here the concept of governance presented as a useful tool for decision-making, nominating, identifying and, when studying a new situation, characterised by a multiplicity of regulatory forms and the fragmentation of power between the various layers that now make up the political-administrative, economic and social reality.

Governance and Urban Policies

In recent years, theories on governance have shaped thinking on urban policies (Dowding 1996, Goldsmith 1997, Le Galès 2006, Stoker 1998). These approaches are primarily concerned with the coordination and merger of public and private resources, which represents the strategy adopted, to a greater or lesser extent, by local authorities across Western Europe. The governance theories seek to aggregate the totality of theoretical concepts around governing and are deemed to be an effective process for orienting and structuring a society (Kooiman 2003, Peters 2002), referring to processes of regulation, coordination and control (Rhodes 1997), analysing processes of coordination and regulation in which the main concern is the role of the government in a governance process understood as an empirical issue (Kooiman 1993, Rhodes 1996, 1997). This also incorporates all types of guidance mechanisms related to public policy processes, involving various types of actors, with a variety of actions associated with different participants and with consequences for governance (Kickert et al. 1997). Public policy is formulated and implemented through a large number of formal and informal institutions, mechanisms and processes commonly referred to as governance (Pierre 2000, Pierre and Peters 2000).

In cities, especially metropolitan areas and what are designated urban regions, new configurations have emerged in the relationships between the state and local power. These alterations in the political and administrative landscape have taken on specific characteristics and dimensions, in which the problems to be resolved are so important to the public authorities that they generate fields in which new forms of public action may be tested but which depend ever more upon contractual negotiations between institutions with diverse and different statutes and purposes (Gaudin 1999). In this institutional and political entanglement driving the major agglomerations, one of the overriding objectives of public policy is to establish scenarios for exchanging and negotiating between institutions staking a claim to legitimately holding part of the general interest and endowed with a proportion of the political resources (technical, financial, juridical, budgetary resources, etc.) essential to all public action. Planning and urban professionals and local authorities have been transformed into nodes in an institutional network. According to certain analysts, this capacity to establish inter-institutional relationships is a crucial resource in the competition between metropolitan areas. The institutionalisation of collective actions seems to represent an issue of equal or greater importance than the spatial framework of the metropolises, and thus the social, political and economic objects are greatly fragmented.

Governance formalised a reconfiguration of the relationships between the institutions and actors participating in the production and implementation of policies applied in metropolises. This largely rests upon processes of restructuring modern states and upon the increasingly important role now attributed to non-state actors and institutions in the regulation of societies.

Within the contexts of the territorial and social fragmentation bound up with processes of internationalisation and the post-industrial nature of cities perceived within the framework of urban governance, we may encounter the opportunities and capacities for urban actors to put policies into practice, in particular for economic development, but also in terms of urban planning, through the capacity to integrate diverse social and political groups and produce shared visions around urban development.

Governance results from the types of 'arrangement' that actors build up, developing a triple capacity for action, integration and adhesion, which results in the capacity for representation. In this sense, a more sociological perception of urban governance is defined. On the one hand, this reflects the capacity to integrate the voicing of local interests, organisations and social groups, and on the other hand, there is the capacity for external representation, to engage in, to a greater or lesser extent, unified relationships with the market, with the state, other cities, and other levels of government.

In terms of the state, the urban governance institutions are themselves restricted by factors such as the organisation of its constitutional and legal conditions and other types of responsibilities ascribed to public organisations. Despite urban governance theories providing a new approach for comparative analyses of urban policies, recognition also needs to be given to the importance of the national context in which this urban governance is enacted. National politics remains a powerful factor for explaining various aspects of urban policies, including the urban economy, urban political conflicts and the strategies in effect for mobilising local resources. The state continues to effectively limit local political choices, remaining the key entity in national sub-affairs (Pierre 2000). The national state is today, more than ever, influential in determining the way in which municipal councils and regions respond to the challenges of globalisation (Harding 1997), as well as the internationalisation of the economy with the state identified as the critical determinant in local political processes (Strom 1996). Analysis of the organisational capacities of local power structures is essential to any understanding of urban management. This happens because these organisations are among the core participants in governing. From the governance perspective, the approach to the core questions is focused upon the role of local power in urban management (Pierre 1999).

Governance and Territorial Management in Metropolitan Areas

These issues take on particular relevance when applied to the government and territorial planning of metropolitan areas in European Union member states associated with the imperatives of competitiveness and cohesion explicit in various community declarations and directives (see Faludi 2006, 2007). As locations concentrating both wealth and knowledge, as well as sources of innovation, the metropolitan areas are crucial actors in economic growth and the future prosperity of European society. They are also generally, and particularly in the cities of Southern Europe, the sites of conflict between the different interests and local powers (municipalities) that tend to polarise development options that might be taken up by inter-municipal and metropolitan structures ending up controlled by the national level, losing capacity for mobilising local and regional interests. It is within this dual debate over the redefinition of competences at the local level, in terms of its interaction with transversal competences at a higher level and within contexts of metropolitan competitiveness, overlapping interests and jurisdictions, that the role of governance gains meaning and pertinence.

The question is explicitly raised in the European Green Book on Territorial Cohesion when referring to the need for cooperation so as to guarantee that EU territorial cohesion objectives are met and when considering how best to proceed with future decisions on community policies – 'Improving territorial cohesion implies better coordination between sectoral and territorial policies and improved coherence between territorial interventions' (Commission of the European Communities 2008).

The concept of territorial cohesion, and consequently the means of guaranteeing its policy objectives, however, is not clear, as may be seen in the results of the public debate promoted by the European Commission.[5]

The differing concerns of entities and institutions representing countries from the north and the south of Europe are renowned. Faludi (2006, 2007) picked up this theme when referring to a 'European model of society' and particularly in the case of France (but extendible to other countries within the 'family' of Napoleonic planning) expressing concerns over a cultural dimension (in contrast to the 'Anglo-Saxon' model) to the territory and its planning supported by a system that implies, for its efficient functioning, an appropriate level of interaction between the differing territorial scopes and levels of responsibility.

> 'Agglomérations' and 'pays' are areas characterized by geographic, economic, cultural or social cohesion, where public and private actors can be mobilized around a territorial project (projet de territoire). There is a link with regulatory planning in that the pays are invited to formulate new-style structure plans, called 'Schéma du cohérence territoriale'. Clearly, in French eyes, the sense of purpose generated by participation in territorial projects is important. (Faludi 2006: 673)

5 http://ec.europa.eu/regional_policy/consultation/terco/contrib_en.htm.

The question lies in the means to establish the conditions for this interaction to take place. In the debate over European Territorial Cohesion, the contribution submitted by the Portuguese Geography Association, itself the result of wide reaching discussion carried out at the national level by this association, made the following appropriate statement:

> Given the finding that the Portuguese legal framework is in itself sufficient, the development of new forms of governance should begin by approaching the lack of coordination between the entities responsible for each sector and between behaviours and the strengthening of actor participation.[6]

This lack of coordination between entities and jurisdictions reflects the tradition of intervention with public decision-makers located at different institutional levels engaged in the same territory. The relationships between the different authorities are essentially based upon shared responsibilities and a division of competences. Meanwhile, the greater autonomy of local power in relation to the state and a European openness promoted changes. There is now a multiplication of contractual relationships between the state and local power and the development of direct relationships between local levels of power and European Union institutions. Thus, the challenge arises out of conciliating national and European priorities and local initiatives and finding new means of interrelating policies enacted across different scales.

Governance and Territorial Management in the Lisbon Metropolitan Area

The case of the Lisbon Metropolitan Area (LMA) is a paradigmatic example of national/regional relational problems, given that there is no elected and representative body at the metropolitan level as there is at the municipal level. This problem is compounded by the prevailing condition of having established a metropolitan area inheriting a non-representative system of government with the absence of a planning structure. That means there was no legal framework in effect for urbanisation processes and projects beyond those undertaken by public initiative, which were furthermore already limited in scope to local councils.

Correspondingly, the growth of Lisbon and its urban region did not happen in accordance with the regulatory models and standards of the majority of European cities. The mode of regulation structuring the interaction between urbanisation and social and economic development was, in the case of Portugal, peripheral and incomplete (Rodrigues 1988). The concern shown regarding city planning as from the late nineteenth century (for example, the expansion with the construction of the *Avenidas Novas*) and the social state intervention in the 1940s, did not extend

6 http://ec.europa.eu/regional_policy/consultation/terco/pdf/4_organisation/134_1_apg_pt.pdf.

out to the wider and more peripheral areas of the city, nor was it then integrated into any coherent social and economic infrastructural model.

Hence, the LMA area experienced urbanisation, intensely and extensively, as a response to the effective demand created by a peripheral model of industrialisation, with processes more illegal than legal, stretching out along the main axes of transport and communication, taking up large areas surrounding the traditional centres through processes allocating plots of land breaking up large estate holdings (Cabral 2004). This process was not complemented by equivalent investment in terms of the conditions for social reproduction given the low role of internal consumption and public infrastructures for economic development (Cabral 2004).

The situation changed significantly from the end of the 1970s with the intense dynamics of urbanisation overwhelming the response capacity of the municipal planning system, representative but still incipient and without any appropriate interrelationship with the decisions handed down by the central administrative entities, in particular the public sector companies responsible for major infrastructures. What happens is not very different to the problems facing many urban regions or metropolitan areas under scenarios of development conditioned by the imperatives for interaction between decision-making levels, actors and institutions with different agendas and priorities and within which the local level is normally the weakest partner.

The relative fragility of the local level is the result of its proximity to users and citizens and a more direct dependence on market dynamics (in land and housing) contrary to the concessionaries and public service providers and higher level political bodies, less immune to conjunctural fluctuations, nevertheless remaining strongly interested in the benefits deriving from higher land values and the visibility and profile generated by viable major urban projects and infrastructural works.

Given its inherent nature defined by functional interdependences and exchanges, the management of the LMA territory, with 19 municipalities, requires a regional administration with effective authority, competences, resources and the legitimacy to tackle and resolve the complex problems facing the region. However, this metropolitan institution, founded in 1991, is endowed with neither the competences nor the resources, partly due to being an unelected body, and has not demonstrated the capacities necessary to deal with the challenges of metropolitan governability.

In this way, governance, given its characteristics, represents a great challenge for strategic modernisation, especially in the case of the LMA, its regions and sub-regions of extensive urban and suburban concentration, with a deficit in territorial planning, excessive state bureaucracy, a lack of coordination of the means available (public and private) and horizontal and vertical interaction, social autonomy, etc.

It is known that between putting forward a specific project/plan and actual implementation, a complex bureaucratic process has to be negotiated, with intermediary decisions, rulings, authorisations, regulations, etc., which frequently combine to induce project delays or even render it non-viable. This all takes place within a universe in which various responsible participants, the state (whether centralised, dispersed or decentralised), civil society, and others, intervene. One of the greatest obstacles to effective governability/governance may be identified in the institutional labyrinth and overlapping competences present within any planning process. One recent study[7] totalled no less than 180 public entities at work in the Lisbon Region, in areas as different as territorial administration, regional development and tourism, among many others.

The Role of Public Administration in Territorial Planning and Development

The results of a research project undertaken by a number of schools of the Technical University of Lisbon on the dynamics of localisation and transformation of the LMA region clarify the importance of the municipal level regarding central state decisions and investments in planning the metropolitan territory.[8]

This study cross-referenced two types of information: the zoning classification proposed in the different Municipal Land Use Plans (*Plano Director Municipal* – PDM) associated with patterns of land use and the built environment registered through to 1992 (the point in time when many PDMs for LMA municipalities came into effect) and between 1992 and 2001. Land use patterns were based on the identification and evaluation of the various classes of utilisation set out in the PDMs resulting from the dominant prevailing land use types.

The existence of various different classifications for land use in LMA municipal PDMs resulted in the development of a methodology for the analysis and definition of common criteria, aggregating land use classifications with similar characteristics. This relative lack of coherence in terms of typology and zoning criteria partially stems from the temporal differences in establishing the plan and the duration of drafting, ratifying and publishing the respective PDMs, which were produced by highly diversified teams.

The various classifications were subject to analysis and comparison as well as their respective constituent criteria. This led to the identification of a set of predominant patterns in accordance with the characteristics of land occupation practices across the LMA. This analysis focused upon an extremely complex reality and it was therefore necessary to carry out aggregations and simplifications to attain the analytical objectives.

The consolidated urban environment corresponds to those territories with a planned and structured dense utilisation of the available urban space. The proposed area for urban development is standardised in all the plans, setting out

---

7 www.gestaoestrategica.ccdr-lvt.pt/1056/estrategia-regional:-lisboa-2020.htm.

8 Project Totta/UTL/01(2004-2008) – *Dinâmicas de Localização, Transformação do Território e Novas Centralidades na Área Metropolitana de Lisboa: que papel para as políticas públicas?* Project financed by the College of Integrated Studies – TUL with the participation of four Technical University of Lisbon – Clara Mendes (FA) (coord.), Romana Xerez (coord. ISCSP), Manuel Brandão Alves (coord. ISEG – CIRIUS), Fernando Nunes da Silva (coord. IST – CESUR) and João Cabral (coord. FA).

large areas for urban expansion defined by the urban perimeters of the various centres, forecasting the strong urban growth identified in the map (Map 2.1).

The areas defined in the PDM as *tecido urbano consolidado* (urban areas) in all municipalities is equal to around 10 per cent of the LMA total area, with the greatest incidence attributed to municipalities on the north side of the Tagus, where this percentage rises to 13.5 per cent. This greater density is connected to the LMA's urbanisation process, which first began around the hinterland of the national capital – Lisbon – supported by the first railway lines to Sintra and Cascais. Regarding the urban areas, the extent of the difference across LMA municipalities ranges from 2.4 per cent in Alcochete, up to the 47.3 per cent in Lisbon.

The filling out of the constructed space in urban areas reaches an average level of around 17 per cent in the LMA, with no major disparities between municipalities. Only two cases stand out, Lisbon with the highest occupancy rate (30.3 per cent) and, at the opposite extreme, Azambuja (4.5 per cent) where the actual built area remains at a residual level.

The areas defined in PDMs as *espaço urbanizável* (urban development areas) across all municipalities account for around 5 per cent of the total LMA area without much difference between the northern and southern municipalities. The highest values, in terms of area, are in Oeiras (26.5 per cent), Almada (38.5 per cent), Barreiro (21.8 per cent) and Amadora (11.4 per cent). On the other hand, we find that the lowest percentage of urban development area is in Lisbon, which is understandable given the level of occupation and consolidation of urban areas.

The density, in terms of area, of buildings in urban areas represents around 5 per cent of the total LMA area, again with little difference between municipalities. The distribution of occupation over the two periods displays some similarity (2.7 per cent up to 1992 and 2.4 per cent between 1992 and 2001). These low levels of urban occupation, recorded up to 2001, call into question the potential and forecast urban growth anticipated by the majority of the PDMs.

Furthermore, even in spaces with restrictions and limitations on construction, such as agricultural, forestry or natural areas, we find some fairly dense incidences of construction where dispersed urbanisation has taken place.

Given these findings, an evaluation of the implementation of this generation of PDM finds, in general terms, an over-scaling of the land classified as urban/urban development areas, with spaces under this classification being far from necessary for urban expansion. This results from the excessively large PDM urban perimeters and extensive areas for urban growth punctuated by disconnected urban development projects. Thus, we encounter here an uncoordinated urban expansion in which land eligible for urban development remains vacant for long periods of time maintaining high land prices, effectively ensuring it stays ineligible for swift intervention by the local authorities.

This overview reflects the role of administrative practises and the institutional framework regulating land use dynamics and changes. Thus, territorial transformations have been poorly influenced by planning as the tools available

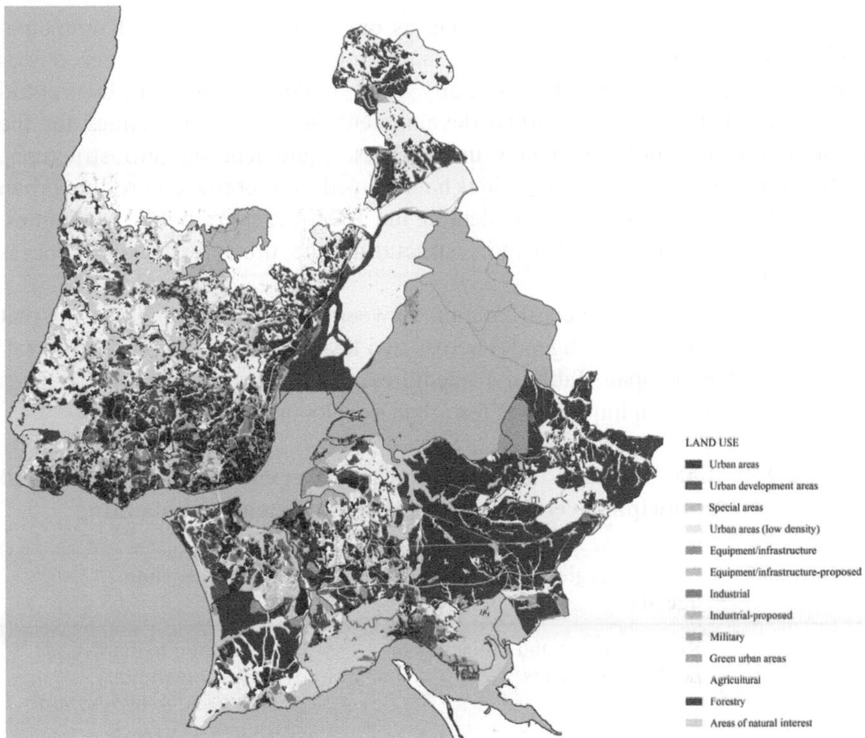

Map 2.1 PDM spatial classifications for the LMA

Source: Project Totta/UTL/01 (2004–2008) – *Dinâmicas de Localização, Transformação do Território e Novas Centralidades na Área Metropolitana de Lisboa: que papel para as políticas públicas?*

have proven non-operational, hence demonstrating the ineffectiveness of traditional rigid planning instruments.

Instead of expressing a model of territorial organisation, the PDMs reflect compromises, driven by the overlapping wishes of central administration and private interests and running counter to the objective of ensuring territorial urban coherence.

The project conclusions highlight the fact that the planning system, based on zoning, had a limited impact in the metropolitan territory. Additionally there was a lack of coordination among the different local systems. One of the aspects that potentially contributed to this lack of coordination and efficiency was the non-operational implementation of a Regional Territorial Plan.[9]

9 At the LMA, tenuous steps have been taken in terms of governance, which are reflected in the relationship established between the CCDR-LVT and the municipal

Another project conclusion was that, as one of the objectives of municipal policies is development control, these have not proven efficient, since it was primarily the infrastructures, especially road building, overseen by central authorities, that conditioned urban development, serving as guidelines for the location and development of new urban areas, equipment and infrastructures. Within this scenario, the municipalities have played a minor role controlling urban development, which became dependent on the effectiveness of public companies, especially roads and infrastructures structuring the urban expansion process (Cabral et al. 2007).

This is the result of the confrontation between two logics of intervention that are associated with distinct agendas across two levels of intervention, the central/regional and municipal. Table 2.1 identifies the most significant differences between agendas with implications for urban development.

Table 2.1 The agendas for the national/regional/metropolitan and municipal levels in territorial management and planning

Dimension	National/regional/metropolitan agenda	Municipal agenda
Political	Rationalisation and efficiency of public investments	Conformation to market dynamics, community pressures and to budget and urban property taxes
Procedural	Legal and institutional framework	Land use control, zoning, public participation
Formal	Coherence and profitability of environmental and infrastructural networks	Conformity between land use, property rights, and public and private interests

Source: Cabral (2006).

Thus, the challenge for metropolitan governance relates to the potential synergies in institutional and operational agendas and to the capacity to foster the interrelationship between the macro–micro scales: strong institutions at the macro level and highly flexible and operational at the micro level (Portas et al. 2003: 39). However, the institutional and operational capacity necessary for the design and implementation of the appropriate policies raises other questions, especially in terms of ensuring compatibility between jurisdictions and the democratic legitimacy of the 'trans-municipal' level, 'altering the approach and

authorities, and specifically in the implementation of cooperation protocols that were agreed based upon the PROT-AML guidelines and their transposition into territorial planning instruments, especially the PDMs.

understanding of the different planning functions, capacity for innovation and additional competences' (Portas et al. 2003: 208).

Flexibility in Local Administration Management: Municipal Companies

The conclusions and data gathered in another recent research project on the LMA highlighted the emergence of institutions, municipal and inter-municipal companies with local government, and management responsibilities with the objective of bringing about greater flexibility and institutional interaction to ensure that services are provided on a financially viable basis. Within this context, local powers have sought to find new ways of managing public assets and interests as a general rule by handing over the direct management to more business-like regimes, seeking out partnerships with other municipalities or other public and private entities. These relationships between municipalities are marked out by broadening the systems of cooperation and exchange and by heightening competition.

The emergence of complementary municipal administrative structures only made sense in the wake of the April 1974 revolution and the 1976 Constitution, with the formalisation of a political power with autonomous administration. Thereafter, the municipalities have been taking on an increasing role in the development of their territories, gradually building up their competences across highly diverse fields.

In Portugal, a process of administrative decentralisation took place in the early 1980s creating the opportunity for Portuguese municipalities to adopt a private sector approach to public services in the early 1990s.[10] The study produced by Crespo (2008) shows how municipal companies have swiftly become the most common form for councils to achieve a diverse range of goals. With new legislation, what was once the exception – that is, companies with public capital replacing municipal departments for the provision of public services – soon became the norm.

At the end of 1999, there were 25 municipal companies on mainland Portugal, with 60 per cent of these concentrated in the two metropolitan areas of Lisbon and Oporto (by 2008, this percentage had dropped by around a half). However, as from 2000, the locations of municipal companies began to expand to take in other municipalities in inland Portugal, particularly in the north of the country.

10 After various attempts, with a range of legislative packages for establishing a legal framework for the founding of municipal companies, Law no. 58/98 was approved enabling such entities to be set up (Silva 2000). This legal framework was revised with the enactment of Law 53-F/2006. The legislation passed under these auspices provided scope for councils to participate in different management models: *Associações de Municípios* (Municipal Associations), *Áreas Metropolitanas e Comunidades Urbanas* (Metropolitan and Community Urban Areas), *Fundações Municipais* (Municipal Foundations), *Sociedades Anónimas* (Limited Companies), whether by shares or cooperatives while municipal companies proved to be the most common option.

In 2001, 269 of the 308 municipalities – thus, 88 per cent of the total – had direct holdings in the capital of public and private companies. Portuguese municipalities held business interests in 114 municipal and inter-municipal companies, 187 limited companies, 58 company shares, 19 banking institutions, 35 cooperatives and 21 foundations – a total of 434 entities.

According to the data available from the General Directorate of Local Authorities, in September 2005, there were 114 municipal and inter-municipal companies. Despite it being legally possible to found municipal companies since 1998, in fact, it was only in 2000 that numbers increased rapidly with the formation of 31 companies (27.2 per cent of a total of 114 companies). In 2001, there was a slight slowdown (28 new companies), however, in conjunction with the previous year, this two-year period accounts for the formation of around 52 per cent of existing companies.

In August 2008, there were 167 municipal companies on mainland Portugal of which 43 per cent are located in the metropolitan areas of Lisbon and Oporto and in the main Portuguese cities corresponding to areas of greatest population. There is no location pattern to these municipal companies, dispersed across the country, except a slightly greater percentage in the north of the country (Map 2.3).

Regarding a classification of the services provided by municipal companies, they are primarily linked to sport, recreation and leisure (27 per cent), culture (20 per cent), tourism (15 per cent) and housing (4 per cent). The idea of municipalities handing over municipal services to municipal companies did not have widespread impact, as shown by the reduced importance of solid waste disposal, water supply and sanitation companies. Municipal companies are mainly associated with the management and maintenance of infrastructures (swimming pools, and theatre, amongst others), particularly in the municipalities outside the main urban areas.

In late 1999, 60 per cent of these municipal companies were concentrated in the two metropolitan areas of Lisbon and Oporto (in 2008, this percentage dropped by about half) with 44 per cent located in the Lisbon Metropolitan Area.

As mentioned above, both nationally and in the LMA, the creation and expansion of municipal companies is associated with the 1988 legislation, resulting in the greatest surge in LMA municipal companies dating to 2000 and 2001. The greatest concentration of municipal companies is in the capital, the Municipality of Lisbon, where the density and diversity of problems is more acute, thus driving the need for new forms of management. The areas of intervention of municipal companies differ across the country but the most common are road transport and infrastructures (Crespo 2008).

The reasons leading to local councils opting to set up a municipal company are not always apparent. Above all, there seem to be casuistic reasons of a political nature[11] with personal attitudes guiding the choice in favour of one model or

11 As shown in Maps 2.4 and 2.5, municipal companies are concentrated among the municipalities on the north bank of the LMA. Municipalities on the south bank are almost exclusively left wing in outlook (Communist Party) contrary to the north bank in which

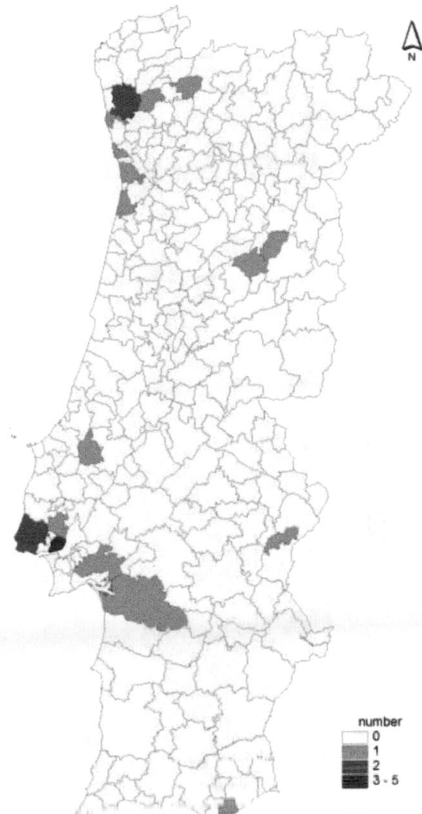

Map 2.2 Municipal companies, Portugal, 1999

Source: Portuguese Association of Municipal Companies, Diário da República, General Directorate of Local Councils, General Inspectorate of Finance.

another, particularly when the entities are set up resulting from the direct initiative of the local authority. The municipal companies may also be established as a type of 'cloaked' privatisation of municipal services. In practice, in the majority of cases, setting up municipal companies meant little more than a transfer of competences of municipal services or departments along with their respective employees, which does not always result in their subsequent abolition.

The juridical status of municipal companies may represent an alternative to direct public management, and this delegated management may result in levels of efficiency equivalent to the private sector. However, this should in no way

---

parties of the centre-left/centre (Socialist Party and Social Democrat Party) predominate. It would seem that it is above all casuistic reasons of a political nature that influence and structure the implementation of new forms of municipal management.

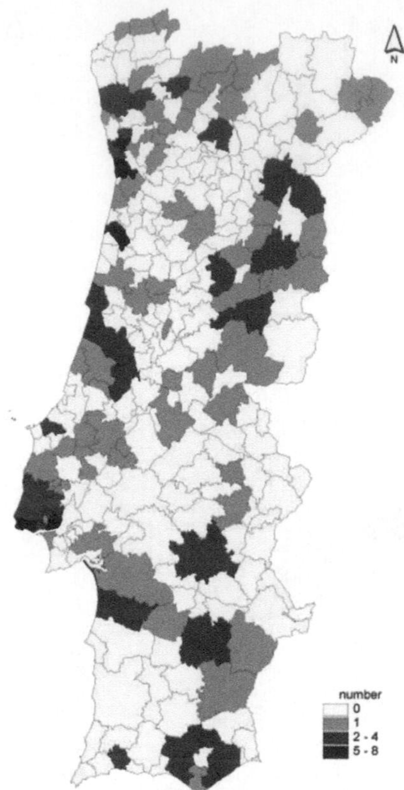

Map 2.3 Municipal companies, Portugal, 2008

Source: Portuguese Association of Municipal Companies, Diário da República, General Directorate of Local Councils, General Inspectorate of Finance.

be perceived as a unique solution for the modernisation of local public services, as there remain countless opportunities for such modernisation to be carried out through the introduction of quality systems into public services. Another observation resulting from the information gathered enables us to state that it is still too early to provide any evaluation of the performance of this solution, especially in comparison with other public management models. For example, the question of the pricing of the services provided is a crucial factor in the decision chosen as regards the different management. In some municipal companies, the tariffs stay above average in comparison with neighbouring municipalities, while in other situations the reverse holds true.

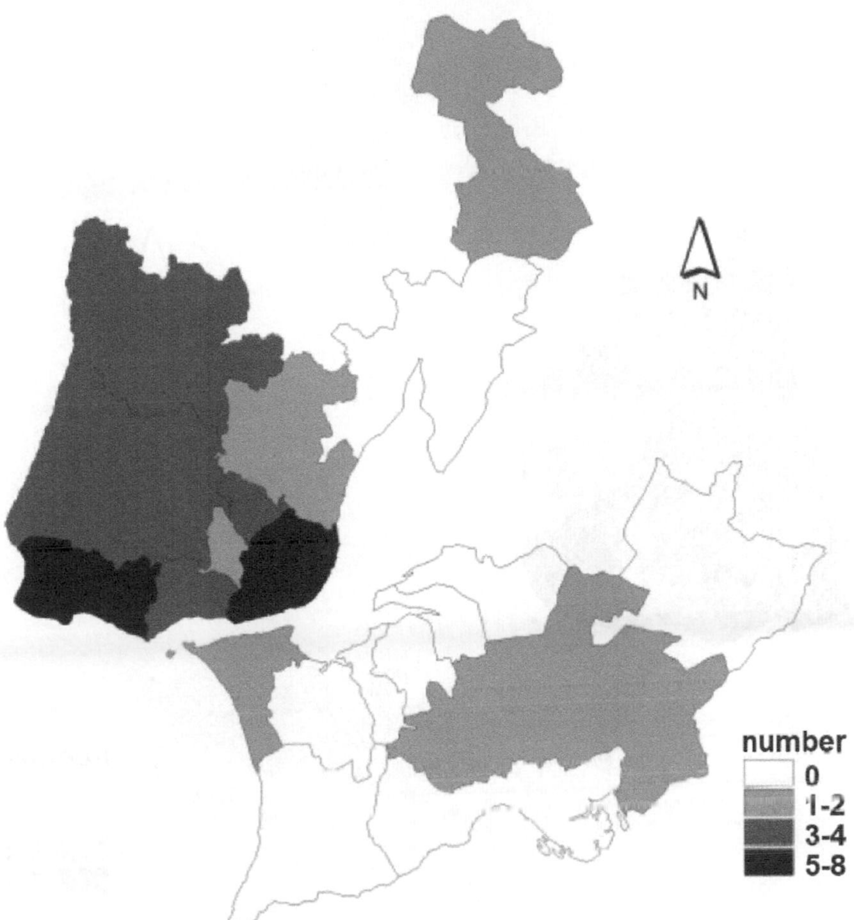

Map 2.4 Municipal companies, LMA, 1999

Source: Portuguese Association of Municipal Companies, Diário da República, General Directorate of Local Councils, General Inspectorate of Finance.

Conclusions

The prevalence of typologies in governance studies reveals a broad spectrum of configurations. The 'governance without government' approach argues that the direction of society on different levels increasingly depends on the interaction of networks of public and private sector actors that are to a large extent beyond the influence and control of central states (Rhodes 1996, 1997). Another approach accepts a central state that retains general primacy in core processes (Gilpin 2001). A third view promotes a more participative style of government although this does

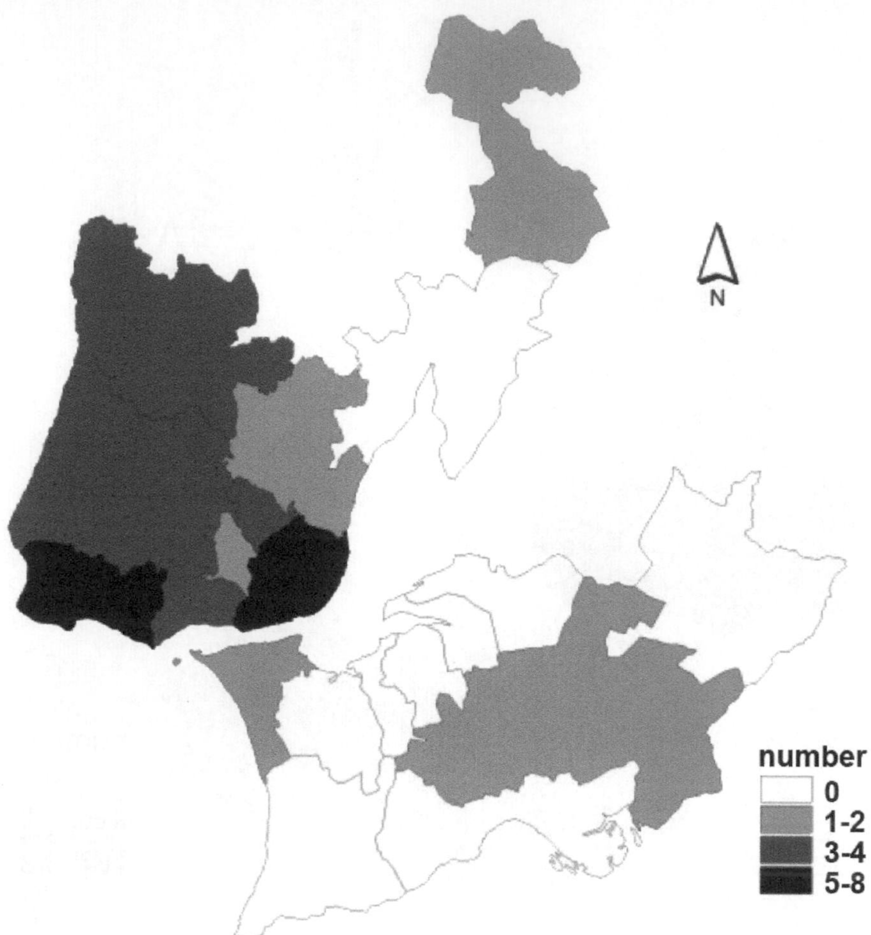

Map 2.5 Municipal companies, LMA, 2008

Source: Portuguese Association of Municipal Companies, Diário da República, General Directorate of Local Councils, General Inspectorate of Finance.

not mean a less powerful government even if in its nominal function – through the concept of steering – the state is restricted to subordinate positions (Kooiman 2003). Hence, the differentiation between government and governance remains somewhat ambiguous, as there are those who affirm that the management of a particular community is 'a responsibility for all members, groups and sectors' (Kooiman 1999) while for others there is a certain susceptibility to absorb governance into conventional aspects of governing. Governance may be inclusive to the extent that it dilutes the border between the two concepts (government/governance).

The changes in the styles of governing involve corresponding changes to the instruments deployed as well as the makeup of the governing class. The changes in the content and targets for government are the clearest transformations. This change in solutions to basic issues became evident during the 1980s and 1990s in the majority of Western European and North American countries, under neo-liberal ideas on the role of the state with a significant reduction in public sector activities.

The governing of territories, traditionally led by the public powers in a centralised and normative fashion, has undergone accelerated mutations induced by the very evolution of society. Indeed, this new role raises serious challenges to the way central administration is internally organised and how it articulates with its surroundings and in terms of restructuring and cooperation, promoting an organisational culture based on dialogue and the surrender of old bureaucratic instruments for control and communication.

Responding to the challenges of contemporary regional (and metropolitan) governance is a necessary condition for the success of the Lisbon 2020 Regional Strategy.[12] However, it is necessary to have a clear understanding of the impact of socio-political processes on the governance of territories in general and to define a framework for institutional interaction in accordance with the administrative and regional civil society realities, as demonstrated by the study on the role of the different levels of intervention in LMA urban transformations, in particular.

In Portugal the opening of the administration to new forms of governance and to the cooperation and participation of the private actors has been rather limited. Thus, the challenge is to promote the involvement and participation of the social and economic actors through partnerships and contracts with the public administration for better governance and efficient management. That requires a more open and transparent institutional environment preventing preferential treatments and overlapping of responsibilities, which means, a) making changes in the functioning of the territorial administration for a clear division of competences and an efficient articulation between the central and local levels; b) to construct and implement strategies and projects with a trans-municipal scope for integration of resources and complementary actions from other sectors of the civil society; and c) to change planning practices and rigid zoning regulations for more flexible procedures capable of addressing and integrating strategic projects.

The challenges to public power require the development and putting into practice new institutional means of intervention. This extends far beyond the replacement of classic modes of public action and political control to include the integration of new procedures, producing new knowledge, and organising the services differently in order to ensure the effective management of public affairs. The emerging role of municipal and inter-municipal companies in the LMA illustrates these trends representing new perspectives on territorial management.

We should above all emphasise that public action should be based upon the alignment of all partners to the territorial project. This involves establishing

12 www.gestaoestrategica.ccdr-lvt.pt/1056/estrategia-regional:-lisboa-2020.htm.

procedures fostering exchanges between all parties, tackling shared problems, progressively building a consensus and putting forward proposals for decision making. Evaluation equally represents a tool of the utmost importance. When effective, this may bring about a better response to the rising complexity of urban politics, strengthen the transparency of public action, inform the opinions of citizens and foster democratic debate.

The notion of governance definitively opened up a field of research that is far from being exhaustively explored. The adaptation of the means of public regulation has continued to undergo change since the beginning of this century. The vocabulary is constantly being enriched with new terms, bringing about the renewal of mental frameworks for an understanding of emerging phenomena. Governance is no longer exclusively or even primarily a mere instrument for strategy, having become an end in itself, a concept for change, and an autonomous doctrine for the practice of modernisation.

References

Bevir, M., Rhodes, R.A.W. and Weller, P. 2003. Comparative governance: prospects and lessons. *Public Administration*, 81(1), 191–210.

Borlini, B. 2004. *Governance e Governance Urbana: Analisi e Definizione del Concetto*. [www.sociologia.unical.it/ais2004/papers/borlini%20paper.pdf].

Cabral, J. 2004. A inovação nas políticas urbanas – Modelos de regulação e sistemas de governança. *GeoINova*, 10, 33–54.

Cabral, J. 2006. Urban development and municipal planning in urban regions – learning from diverse planning and governance cultures and systems. *WPSC-06 World Planning Schools Congress*. Mexico City, Mexico, 15 July.

Cabral, J., Morgado, S., Crespo, J. and Coelho, C. 2007. Urbanisation trends and urban planning in the Lisbon Metropolitan Area, in *A Portrait of State-of-the-Art Research at the Technical University of Lisbon*, edited by M. Pereira. New York: Springer, 557–72.

Commission of the European Communities. 2008. *Green Paper on Territorial Cohesion Turning Territorial Diversity into Strength*. Luxembourg: Office for Official Publications of the European Communities.

Crespo, J. 2008. Urban management in a governance context. The 'municipal companies' in the Lisbon Metropolitan Area'. *XI EURA Conference – Learning Cities in a Knowledge Based Society*. Milan, Italy.

Dowding, K. 1996. Public choice and local governance, in *Rethinking Local Democracy*, edited by D. King and G. Stoker. London: Macmillan, 50–66.

Faludi, A. 2006. From European spatial development to territorial cohesion policy. *Regional Studies*, 40(6), 667–78.

Faludi, A. 2007. The European model of society, in *Territorial Cohesion and the European Model of Society*, edited by A. Faludi. Cambridge: Lincoln Institute of Land Policy, 1–22.

Gaudin, J.P. 1999. *Gouverner par Contrat: l'Action Publique en Question*. Paris: Presses de Sciences Po.

Gilpin, R. 2001. *Global Political Economy: Understanding the International Economic Order*. Princeton: Princeton University Press.

Goldsmith, M. 1997. Changing patterns of local government. *ECPR News*, 9, 6–7.

Harding, A. 1997. Urban regimes in a Europe of the cities? *European Urban and Regional Studies*, 4(4), 291–314.

Jessop, B. 1998. The rise of governance and the risks of failure: the case of economic development. *International Social Science Journal*/UNESCO, 50, 29–45.

Jordan, A., Wurzel, R.K.W. and Zito, A. 2005. The rise of 'new' policy instruments in comparative perspective: has governance eclipsed government? *Political Studies*, 53(3), 477–96.

Kickert, W.J.M., Klijn, E.H. and Koppenjan, J.F.M. 1997. *Managing Complex Networks: Strategies for the Public Sector*. London: Sage.

Kjær, A. 2004. *Governance. Key Concepts*. Cambridge: Policy Press.

Kooiman, J. 1993. *Modern Governance: New Government-Society Interactions*. London: Sage.

Kooiman, J. 1999. Social-political governance: overview, reflections and design. *Public Management*, 1(1), 67–92.

Kooiman, J. 2003. *Governing as Governance*. London: Sage.

Le Galès, P. 1995. Du gouvernment des villes à la gouvernance urbaine. *Revue Française de Science Politique*, 45(1), 57–95.

Le Galès, P. 2003. *Le Retour des Villes Européennes: Sociétés Urbaines, Mondialisation, Gouvernement et Gouvernance*. Paris: Presses de Sciences Po.

Le Galès, P. 2006. *Gouvernement et Gouvernance des Territoires*. Paris: La Documentation Française.

Leresche, J.P. 2002. *Gouvernance Locale, Coopération et Légitimité. Le Cas Suisse dans une Perspective Comparée*. Paris: Pedone.

Milward, B. 2004. *The Potential and Problems of Democratic Network Governance. Democratic Network Governance*. Copenhagen: Centre for Democratic Network Governance.

Peters, B.G. 2002. *Governance: A Garbage Can Perspective*. Wien: Institut für Höhere Studien.

Peters, B.G. and Pierre, J. 2000. Citizens versus the new public manager. The problem of mutual empowerment. *Administration & Society*, 32(1), 9–28.

Peters, B.G. and Pierre, J. 2003. *Handbook of Public Administration*. London: Sage.

Pierre, J. 1999. Models of urban governance. The institutional dimension of urban politics. *Urban Affairs Review*, 34(3), 372–96.

Pierre, J. 2000. *Debating Governance: Authority, Steering, and Democracy*. Oxford: Oxford University Press.

Pierre, J. 2005. Comparative urban governance. Uncovering complex causalities. *Urban Affairs Review*, 40(4), 446–62.

Pierre, J. and Peters, B.G. 2000. *Governance, Politics and the State*. Basingstoke: Macmillan.

Pollitt, C. 2000. Is the emperor in his underwear? An analysis of the impacts of public management reform. *Public Management*, 2(2), 181–99.

Portas, N., Domingues, A. and Cabral, J. 2003. *Políticas Urbanas – Tendências, Estratégias e Oportunidades*. Lisbon: Calouste Gulbenkian Foundation.

Rhodes, R.A.W. 1996. The new governance: governing without government. *Policy Studies*, XLIV, 652–67.

Rhodes, R.A.W. 1997. *Understanding Governance: Policy Network, Governance, Reflexivity and Accountability*. Berkshire: Open University Press.

Rodrigues, M.J. 1988. *O Sistema de Emprego em Portugal*. Lisbon: Dom Quixote.

Scharpf, F. 2001. Notes toward a theory of multilevel governing in Europe. *Scandinavian Political Studies*, 24(1), 1–26.

Silva, C. 2000. Empresas Municipais – Um exemplo de inovação em gestão urbana. *Cadernos Municipais – Revista de Acção Regional e Local*, XIV(69/70), 13–20.

Sorensen, E. 2002. Democratic theory and network governance. *Administrative Theory & Praxis*, 24(4), 693–720.

Sorensen, E. 2005. Metagovernance: the changing role of politicians in processes of democratic governance. *The American Review of Public Administration*, 36(1), 98–114.

Sorensen, E. and Torfing, J. 2003. Network politics, political capital, and democracy. *International Journal of Public Administration*, 26(6), 609–34.

Stoker, G. 1998. Cinq propositions pour une théorie de la gouvernance. *Revue International des Sciences Sociales*, 155, 19–30.

Strom, E. 1996. In search of the growth coalition: American urban theories and the redevelopment of Berlin. *Urban Affairs Review*, 31, 455–81.

Chapter 3
Competitiveness and Cohesion: Urban Government and Governance's Strains of Italian Cities[1]

Francesca Governa[2]

Introduction

Changes of the organisational forms and modes of action of the Nation-State give rise to two simultaneous levels of redefinition: the redefinition of the national territory as a stable framework of reference and belonging and the redefinition of the role of the State as a political entity in charge of regulation and redistribution (Cassese 2001). According to Jessop (1994), these processes involve, at least partially, a hollowing-out of the Nation-State with the consequent re-articulation of powers and responsibilities towards supra- and infra-national institutional levels (e.g. European Union and local authorities) and towards horizontal networks of power, acting independently of the institutional processes of functions and competencies decentralisation. Changes of roles and functions do not occur separately from spatial changes: the redefinition of State functions is accompanied, and simultaneously intensified, by the re-scaling of State territoriality (Brenner 1999, 2004). Therefore, supra- and infra-national territories are called upon to play a new economic, social, symbolic and political role, with the consequent multiplication of spatial subdivisions and of the stages for policies and interventions; a diversification of scales within which collective actions attain their relevance; and a proliferation of conflicts of both representation and power (Debarbieux and Vanier 2002, Vanier 1999). The change in the territorial organisation of the State thus intersects with the transformation of its roles and of the forms of national statehood (Rhodes 2000). The focus on these aspects overcomes a conception of

1 I would like to thank Giuseppe Dematteis, Matteo Bolocan Goldstein and Gabriele Pasqui for reading an early version of this chapter and providing both valuable feedback and encouragement. I would also like to thank the editors of the book for their supportive comments. The usual disclaimers apply. An early version of this text was published in 2010 in *Análise Social. Revista do Instituto de Ciências Sociais da Universidade de Lisboa*, 197, 663–83.

2 Politecnico e Università di Torino; Interuniversity Department of Regional and Urban Studies. Viale Mattioli, 39; 10124 Turin (Italy). francesca.governa@polito.it.

the State as mere 'container' of organised hierarchical power, highlighting the problem of transversal interaction and coordination between different levels of territorial organisations and of the political and institutional action (Brenner 2000).

However, albeit spatially reconfigured, Nation-State institutions do not passively undergo such processes, but engage in them as actors in their own right. In this sense they continue to play a key role in the political and economical restructuration at all geographical levels, as well as to formulate, implement, coordinate and supervise policies. The redefinition of both the institutional levels and the areas in which political and institutional actions shape themselves

> has not generated a unidirectional process on a single scale – be it European, regional, or local – is replacing the national scale as the primary level of political and economic coordination. (Brenner 2004: 3)

New modes of public action thus define a shift in the meaning of the role of the State and they do not represent evidence of its decline, overcoming the poor and in many ways controversial dichotomous opposition between government and governance. Indeed, the distinction between these two models is not entirely clear, and refers to a *continuum* of intersecting aspects and features of both government and governance. From this perspective, the emergence of governance

> should … not be taken as proof of the decline of the state but rather of the state's ability to adapt to external changes. Indeed … governance as it emerged during the 1990s could be seen as institutional responses to rapid changes in the state's environment. (Pierre 2000a: 3)

Within this framework, the chapter will present and discuss the characteristic of the Italian urban governance, from both the conceptual and the empirical point of view, with the aim of showing its limitations and opportunities. Since the 1990s, Italy has been going through a period of intense change concerning its political and institutional system. As in other European countries more or less during the same period, the beginning of the decentralisation processes has led to a reshaping of the relationships between the State and the local governments, establishing a new framework of competences and defining a new model for the public action (Cammelli 2007). The change of the organisation and the role of the State (and more generally of the public actor) also affects a change of urban government, spreading the (ambiguous, problematic and often purely rhetorical) term of urban governance (Balducci 2000, Bolocan Goldstein 2000, Palermo 2009, Perulli 2004).[3] Italian urban policies are therefore called upon to deal with the changes

3 The ambiguity and the risk of a purely rhetorical or instrumental use of the term governance is a baseline datum (see Governa 2004, Governa et al. 2009, Osmont 1998, Pierre 2000b, Rhodes 1997). Suffice to recall here that often, in more simplistic and streamlined interpretations, the term governance is used to indicate a (generic) improvement

identified by the international debate in urban governance (Davoudi et al. 2008): inter-institutional forms of cooperation between various levels of government; coordination between a multiplicity of actors and interests; involvement of private sector's institutions, and of direct participation of citizens in the decision-making processes. Reading urban policy documents prepared by various Italian cities during this period reveals a clear change in the objectives and forms of the public action. Much less clear is whether, and how, the different dimensions of the urban governance have gained real influence even in the governance practices of Italian cities and in the capacity of policies to address urban problems.

The chapter is structured as follows: after setting out all the legislative and institutional changes that form the framework of the existing urban policies in Italy, it will present the main plans and programmes of urban governance, to discuss then some critical nodes of the Italian experience. These nodes refer to the emergence of a change which is rather more said than done, characterised by inertias against innovation, the resurfacing of old problems, and the difficult 'balancing' between patterns of public action seeking, at least theoretically, to 'hold together' the needs for economic development and the well-being of the population.

The Italy of the 1990s: Decentralisation Process and Institutional Changes for the Urban Government

The plot of the relationships among the processes of political and institutional decentralisation, the redefinition of the relationships between the State, the local authorities, and the urban and spatial policies may be identified by focusing on the forms and the modality of cities' government, beginning with the identification of the 'new' instruments of public policies.[4]

Gathering the different aspects of innovation is not an easy task. Innovation is indeed often only 'superficial' and frequently ending in the spreading of new slogans which conceal very traditional behavioural styles concerning the operative modus operandi of the public administration and of the major economical and

---

in relation to the past (in critical terms, for example, Imrie and Raco 1999). The word governance is thus used as a *panacea* to solve every problem, without taking into account the many significances that are layered around the term, the many meanings with which it has been (and still is) used, the many fields of the exercise of government where governance models are used (Rhodes 1997). This attitude also tends to favour a prescriptive-normative conception of the term (which emphasises how we should act in a certain situation or, more generally, to change the style of government) rather than analytical-descriptive (describing a particular situation or a particular government practice) (Bevir 2002).

4 For an interpretation of urban policies as public policies that, as such, exist and become enacted through a selection of political instruments, see Lascoumes and Le Galès (2004).

social actors. Moreover, innovation does not always directly improves the state of things; facets of innovation are mixed with the preservation of 'old' organisational structures and the inevitable inertia that lurks in the processes of design and implementation of policies (Governa 2004). Finally, new instruments do not always correspond to new practices from public administrations, neither to different modes of urban action or to more efficient and effective outcomes of urban issues (Palermo 2009).

Despite these limitations, there is evidence of the path followed by the changes that have occurred in the context of Italian urban policies. They correspond to the main innovations in legislation that triggers changes in urban governance (Cammelli 2007, 2011, Vandelli 2000), and in particular to the:

- L. 8 June 1990, No. 142 (*Ordinamento delle autonomie locali*): it sets the key principles for the organisation of local authorities and determines their functions. The law also sets the establishment of metropolitan areas (never implemented), defining as metropolitan areas the urban agglomeration of Turin, Milan, Venice, Genoa, Bologna, Florence, Rome, Bari and Naples.
- L. 25 March 1993, No. 81: it changes the electoral mechanism at the local level, with the introduction of the direct election of Mayors, Provincial Presidents and Provincial and Municipal Councils (supplemented by L. April 30, 1999, No. 120).
- L. 23 December 1996, No. 662: it introduces measures to rationalise public finances and instruments for negotiation between public institutions and public and private interests.
- L. 15 March 1997, No. 59 (*Delega al governo per il conferimento di funzioni e competenze alle regioni ed enti locali, per la riforma della Pubblica Amministrazione e per la semplificazione amministrativa*) that, referring to the subsidiary principle, devolves administrative functions and duties for the promotion of local and regional development to regions and local authorities.
- L. 15 March 1997, No. 127 (*Misure urgenti per lo snellimento dell'attività amministrativa e dei procedimenti di decisione e controllo*) and Decree Law of May 31, 1998, No. 11 (*Conferimento di funzioni e compiti amministrativi dello Stato alle regioni e agli enti locali, in attuazione del capo* 1 della L. No. 59) (the so-called Bassanini reform, named after the Minister of Public Administration): they redraw the distribution of powers and functions between the State, regions and local authorities, ascribing to both regional and local levels a very wide range of functions and powers.
- L. 3 August 1999, No. 265 (*Disposizioni in material di autonomia e ordinamento degli enti locali, nonché modifiche della* L. 142/90): it modifies and rationalises the general framework of the previous law, considering both the changes that occurred and the implementation difficulties, particularly with regard to the metropolitan cities and their delimitation. The metropolitan city should incorporated those government

powers attributed to provinces as already set out by the L. 142/90. In fact, even this legal mechanism was never implemented.

The movements towards decentralisation and the progressive local government reform date back to the mid-1980s, with a gradual acceleration in the early 1990s, at a time of deep crisis in the Italian political and institutional system. Measures taken gain the form of an attempt to radically change the institutional arrangements, reforming the systems of control and distribution of responsibilities and power between State and local authorities. The attempt was to simplify the administrative interventions and boost public administration efficiency. The principles from which these laws take inspiration fit in with the trend towards the progressive decentralisation of administrative and political action, which concerns (at roughly about the same time) all European countries (Governa et al. 2009). Such principles seem to envisage, at least in general terms, the redesign of the relationship between State, local authorities and civil society following an entirely different approach from the traditionally institutional arrangements of the Western world, which Bruno Dente (1999: 113) summarised in the formula of the 'riduzione amministrativa della complessità' [administrative reduction of complexity], namely of 'razionalizzazione via accentramento e riduzione dei conflitti attraverso la loro sussunzione a livello superiore' [rationalisation through centralisation, and reduction of conflicts by their subsumption at higher level].

This phase ended, not only symbolically, with the amendment of the Italian Republic Constitution of 1948. Such change, implemented through the constitutional law No. 1 of 1999 and No. 3 of 2001 and the subsequent confirmative referendum of 7 October 2001, covers nine Constitution Articles, contained in Title V, concerning the territorial ruling of the State. Basically, the modification of the Title V sought to establish the foundations and preconditions for a future transformation of Italy into a federal republic based on the principles of subsidiarity (vertical, e.g. between different levels of government, and horizontal, e.g. between public authorities and citizens), and on the announcements of legislative, administrative and fiscal federalism. To the State are essentially reserved exclusive legislative powers only in certain fields (such as foreign policy and international relations, immigration, defence and security), while issues of urban and regional planning (now called *governo del territorio*), of economic development, of scientific and technological research, of transport and communications, and of cultural heritage and environmental preservation are matters of shared legislative competence with the regional level.[5] The administrative functions, even concerning matter reserved to the national level in terms of legislation, are allocated across the various levels of government according to an ascending criterion (often termed as vertical subsidiarity), which favours the local level and, *in primis*, the municipalities. Article 114 also details that the Italian Republic is formed of municipalities,

5 On the meaning of the expression 'governo del territorio' in the Italian debate, even before its introduction in the reform of the Italian Constitution, see Mazza (2011).

provinces, metropolitan cities, regions and the State. All are autonomous entities with their own respective statutes, powers and functions. For example, the metropolitan cities, although being an administrative level never implemented in Italy despite a constant recall of its necessity since roughly two decades, extend beyond the ordinary legal framework to enter directly into the Constitution.

The reforms that led to a progressive increase in the administrative and institutional competences of regions and local authorities, beginning with the administrative decentralisation and the reform of Constitution's Title V, however, were not accompanied by any simultaneous process of recognition of autonomy and financial accountability at the local level. In parallel, the transfer of central powers and functions to municipalities and the need for the local authorities to contain services' costs, while still preserving their quality, have led to the introduction of new forms of public service management. Market forms or mixed forms of 'quasi market' in the management of local services has emerged, also through the adoption of private sector working methods and organisation. The increase in the outsourcing of most local public services, firstly those related to economic activities and then also those related to human beings, has progressively weakened the political and technical role of local authorities. Therefore, within this context the reference to models of governance has been basically translated into the adoption of the principles of the minimal State (Rhodes 2000).[6] Overall, the 'time lag' between the devolution of functions, competences and resources and the gradual 'retreat' of the role of public bodies in the delivery of public services have led to the revival of a centre/periphery model in the relationship between the State and local authorities, which is clearly expressed in the fiscal side of the federalist proposals enacted into law in May 2009. In practice, the central government, more than just retaining regulatory powers, the ability to propose interventions and the determinant resources for their implementation, also maintains a strong influence both on the side of revenues and real taxation, and in terms of the spending and its relative functions. The local level, on the other hand, appears paradoxically undermined from both the legal and financial perspective, and operates within the narrow channel between the sharing of central powers and attempts to defend its own options.

Asymmetries in relations between State and local authorities are complicated by the traditional underestimation of the 'urban' specificity from an institutional point of view. In fact, within the Italian institutional framework, the 'city' is simply a 'larger municipality' and even into the most recent laws (e.g. the Decree Law N. 267, 18 August 2000) the 'label' of city is only a purely honorary act (Cammelli 2011).

The framework of Italian urban governance has also been influenced by the role played by the European Union (EU).[7] Initiatives and community programmes have

6 In the framework of the minimal State, according to Stoker (1998: 18), governance could be trivially understood only as 'the acceptable face of spending cuts'.

7 According to Faludi and Janin Rivolin (2005), the specific style of policy-making of Southern European countries is a key element in building a 'Mediterranean way' to

indeed spread in the Italian practices the 'core principles' and the current European urban policy's mainstream (Janin Rivolin 2003, Gualini 2004). In particular, the 1999 ESDP (European Spatial Development Perspective), INTERREG and URBAN initiatives, the regional policy and procedures for managing structural funds, and the *Territorial agenda* of May 2007 have together gradually changed, in different but converging manners, the Italian public administration action. It might be enough to consider, as an example, the 'revolution' introduced in the Italian practices by the widespread adoption of the competitive procedure for the allocation of financial resources, with the consequential diffusion of evaluation procedures and rewarding mechanisms. At the same time, local authority autonomy, subsidiarity, accountability, appropriateness of the public structures to the carrying out of the responsibilities assigned to them, flexibility in inter-institutional relationships, citizen participation in collective choices, and streamlining the bureaucracy, i.e. the key principles of EU policy approach (see Janin Rivolin 2003), became, even in Italy, the cardinal points that structure urban policies.

The relevance of this set of changes, beyond declarations of intent, should obviously be tested in practice in order to assess whether and how they affected the attitudes and styles of public administration's government of Italian cities.

The Change in the Practices: Projects and Programmes of Urban Governance

Institutional and legislative changes 'translate' into a variety of policies, plans and interventions of different origins and natures, which have changed the government of Italian cities, spreading principles of urban governance rapidly, although often only superficially. Amongst the many projects and plans that have changed, and continue to change, the way in which Italian cities are managed and governed, main examples of the characteristics assumed by Italian urban governance are the strategic plans, whose overriding goal is to promote the urban competitiveness in the global arena, and the so-called complex programmes, aimed at urban regeneration. The reference to the dimensions of governance (from horizontal and vertical subsidiarity to citizens participation, from the territorialisation of policies to the changing role of public body, see Davoudi et al. 2008), seems particularly evident, at least in intentions, in these two areas of urban government. However, many of the transformations which took place in Italian cities in recent years, and their forms of government, are often the result of decisions, actions and policies made outside of the governance framework. These include sectoral policies, in particular those relating to public transport and mobility; changes in the local service management systems, with a rising hollowing-out of the political and technical role of municipalities; the implementation of administrative decentralisation, with the progressive increase of the powers of regions and

---

European policies.

local authorities, accompanied by a steady reduction in the transfer of financial resources and the continuing lack of recognition of financial independence. These are all examples of processes which have deeply affected Italian cities, on which there has never been a systematic effort to interpret the various aspects of urban governance, on any sort of coherent or strategic framework.[8]

Within the programmes that explicitly refer to the terms of governance, the examples given by the strategic planning, as emerging from the not so many critical reflections on the Italian experience (e.g. Palermo 2009, Perulli 2004), enable us to detect a specific trend of the urban governance in Italy. Italian strategic plans fit into a conception of urban governance outlining the city government's ability to successfully confront the diversity and fragmentation of actors and interests that act in the decision-making arena to promote competitiveness and economic development. The assumption underlying this interpretation is that the city can be interpreted and managed as a 'collective actor' (Bagnasco and Le Galès 1997, Le Galès 2002). Hence, the strategic plans become a tool that assumes and expresses the formation of the 'city's collective actor'. The aim is to define a shared vision of the urban future within which are manifested and integrated the interests of a number of actors and social groups, in order to define the positioning strategies of the city towards the market, the State, and other cities considered to be, according to the dominant rhetoric, benchmarks (see Conti 2002).

In Italy strategic planning was established in the mid-1990s, relatively late compared with other European countries (Barcelona, considered a 'beacon' for the Italian strategic plans, began this in 1988). The first Italian city to adopt a strategic plan was Turin in 2000, with a plan subsequently revised and updated in 2006 (Torino Internazionale 2000, 2006). However, Italian cities have recovered quite quickly the gap from the slow start. The strategic plan of Turin was quickly followed by the experiences of La Spezia, Florence, Cagliari, Asti, Bari, Bolzano, Carbonia, Jesi, Prato, Venice, Naples and others. Hence, the model of strategic planning, thanks also to the funding set up by the Ministry of Infrastructures and Transport, spreads throughout the country, from north to south, and in cities of varying sizes and with different development opportunities and issues. Many Italian cities have turned to the strategic plan as a tool to address problems posed by the crisis of the old industrial model and by the need to promote the local economy and employment, especially so in the light of the progressive decline in the redistributive capacities of the central state.

Obviously, it is difficult to make an overall assessment of the strategic plans of the various Italian cities, as each represents a special case. However, reflecting again on the more general trends than on the individual cases, it is still possible to identify

8 The role of these processes in the transformation of Italian cities has not been thoroughly analysed for now. On the relationship between constitutional reform and local system of social services, see Tubertini (2007); for a general assessment on the transformations of the mayor Italian cities – Turin, Milan, Rome and Naples – during the period 1993–2010, see Amato et al. (2011).

the main strengths and weaknesses that characterise the Italian strategic planning experience. The major factor of interest is the reference to a strategic dimension of the urban government, which points to a change of Italian practices long centred on a strictly technical view of planning. According to Palermo (2009: 113):

> Dopo aver esplorato modelli improbabili di governo del territorio, mediante piani prescrittivi di notevole grado di dettaglio ed esteso orizzonte temporale (una singolarità italiana); dopo aver sperimentato, con decenni di ritardi ed esiti controversi, I modelli della programmazione strutturale e della progettazione integrate che dispongono di esperienze mature in altri contesti evoluti; or ail nostro paese si dedica più volentieri ai più agevoli esercizi di pianificazione strategica, che per definizione implicano minori responsabilità di scelta e di azione. [After having explored unlikely models of *governo del territorio* through prescriptive plans of a considerable degree of detail and extended time horizon (an Italian singularity); after having experimented, with decades of delays and controversial results, the structural models of planning and integrated programming who have interesting experiences in other contexts; now our country is willing to spend on strategic planning exercises, which, by definition, involve minor responsibilities in choice and action].

The strategic content of the urban government puts at the heart of the planning process also in Italy the role of the political and social dynamics of cities, the power relationships present within them, and the supra-local flows and relationships in which cities are included. Meanwhile, Italian strategic plans are often configured around a joint set of generic objectives, as purely rhetorical statements unable to express an effective capacity for action, or even as documents that promote, rather than the elusive 'city's collective actor', the development strategy expressed by the dominant urban regimes and coalitions of interest. Italian strategic plans are therefore frequently constructed as a set of goals entirely alien to the project of 'physical' transformation of the city, thanks to the failure of the relationship between the definition of strategies and the operative choices of urban planning, and also to the (often) lack of direct mobilisation of responsibilities by the actors involved in the project implementation. Conversely, when strategic choices rely physically on 'big urban projects', relating to the transformation of significant sections of the urban fabric, the strategic plans are 'crushed' under the interests of ruling elites and, in particular, under those with interests in the building and construction sectors, ending mainly in mere real estate operations.

The role of urban regimes and coalitions of interest that govern cities is not, of course, a novelty. It might be sufficient to take into consideration the wide international literature on the city as a growth machine and on the urban regime (Dowding 2001, Logan and Molotch 1987, Molotch 1976, Mossberger 1994, Mossberger and Stoker 2001, Stoker and Lauria 1996, Stone 2005). What is surprising in the Italian case is that the strategic plans are often only vague attempts to assert the existence of an aggregate dimension of interests and that, even more frequently, it is difficult to perceive the presence of a different strategic vision of urban government (and of the city itself). In essence, what emerges from the Italian strategic plans is the difficulty of the urban elites to fully express their interests and to make coalitions, to mobilise the necessary resources towards urban development, and to direct strategic choices towards a new spatial order (Mazza 2000). Even the strategic plan of Turin, for a long time listed as a 'virtuous example' of the Italian strategic plans, has been more recently subject to a critical interpretation that emphasised its rhetorical and highly ornamental dimensions (Palermo 2009; on Turin's coalitions of interest, and their difficulties, see Belligni et al. 2008).

The second set of projects where it is possible to find 'traces' of urban governance in Italy are the so-called complex urban programmes, which constitute the 'Italian way' to urban regeneration (Governa and Saccomani 2004). These programmes, which have often been implemented in accordance with EU URBAN initiative or through projects emulating its 'spirit', have disseminated urban governance practices based on the integration principle, inter-sectoral approach, and methodologies of participation of inhabitants.

The complex urban programmes have had a widespread deployment and have been used as means of intervention for urban regeneration in many Italian cities of varying sizes and importance from both the economic and territorial perspective. In addition to regional capitals, which have witnessed the problems of exclusion and marginalisation typical of major cities (Milan, Turin, Rome, Naples, Palermo), experiences of urban regeneration have also been implemented in small towns (Cosenza, Foggia, Livorno, Rovigo, Salerno, Savona, Syracuse) or in smaller municipalities, often located in suburban areas near large cities (Seregno and Cinisello Balsamo, near Milan, or Settimo Torinese and Venaria Reale, near Turin).

It is difficult to determine the results actually achieved by such programmes because they are locally highly differentiated. However, despite the obvious differences, some elements characterise all Italian urban regeneration practices (Governa and Saccomani 2004).

The complex urban programmes have changed the way in which urban regeneration was established in Italian cities, traditionally focused on physical interventions. The role of EU URBAN initiative, or the inspiration from its guide-principles, allowed channelling measures into relatively small areas within cities, to integrate social, environmental and economic initiatives and to spread methods and techniques for the direct participation of citizens in decision-making processes. In this sense, urban governance in Italy has experienced various forms of integration between sector and policies, partnerships, definition of financial frameworks, as well as less bureaucratic relations with urban planning instruments and legislation.

There are also critical issues. In the Italian urban regeneration programmes, inequalities and the difficulties encountered by the cities are deemed local problems requiring local solutions. This approach is based on a specific approach to describe and interpret the city (see, for a critical view, Amin and Thrift 2002): the city is seen and conceptualised as a defined and static space, somehow 'waterproof' to external flows and relations. Through forms of 'local mobilisation', interventions implemented in urban regeneration programmes seek to promote the social *mixité* or to improve the built environment of neighbourhoods but do not try to 'fight' the supra-local sources of injustice. Hence the *ratio* of urban regeneration programmes seems to fit perfectly with the 'celebration' of the local (Governa 2008), i.e. the a priori assumption that local actions are preferable to others, in relation to the supposed or real capability of the local level to develop projects and strategies that are more effective, democratic, sustainable and right (Purcell 2006).

The issues of integration and participation, which are those of great potential innovation in international experiences of urban regeneration (e.g. Atkinson 2000, Carpenter 2006, Chorianopoulos 2002, Parkinson 1998), further define clear limits to the Italian practices of urban regeneration. Despite the premise and attempts to integrate different activities and policies, in practice, interventions appear often focused on issues related to estate and physical regeneration of building and neighbourhoods. The difficulties to manage the participation of the multiplicity of actors and interests of the urban regeneration arena are also clear. Furthermore, it is not only in Italy that the issue of participation is often used as a 'rhetorical weapon' to include intervention methods challenged by dual, and in many ways, contradictory requirements: on one hand, the involvement of the 'strong' actors (investors, entrepreneurs), which guarantee the funding of initiatives; on the other hand, the participation 'from below', which supports the promotion of social cohesion and citizen empowerment (Geddes 2000, Tosi 1994). In fact, the most common form of participation concerns the organised interests (public and private) and rarely draws on the diffuse involvement of the population, although it is often possible to observe public consultations to inform the citizen on decisions made. However, in the experiences of Italian regeneration, private actors do not seem to have found sufficient reasons to generate suitable investments to start a virtuous circle halting and reversing processes of degradation through better training and employment. On the other hand, 'inclusive actions', designed to promote widespread citizen participation in decision-making process, present both difficulties and limitations (see Regonini 2005). Not only in Italy, but perhaps more in Italy than elsewhere, participatory practices are full of 'rhetoric and craftiness' (Paba 2003), often referring to purely ritual forms, already stigmatised in the late 1960s by Sherry Arnstein; to 'virtuous' practices only feasible when it comes to micro-decisions relating to severely limited objectives and goals (and therefore not meeting the most important interests) or even to 'resistence' strategies towards projects or interventions.

Italy's Urban Governance Myths and Illusions

Overall, the political institutional decentralisation process in Italy, and the innovations it gives rise to, seems to oscillate between the two 'forms of devolution' described by Hudson (2005). According to this author, in fact,

> what is claimed to be new and qualitatively different about more recent regional devolution is that it encompasses the power to decide, plus resources to implement decisions, at the regional level. Others, however, dispute this, and argue that what has been devolved to the regional level is responsibility without authority, power and resources. (Hudson 2005: 620–1)

In Italy, as elsewhere, the decentralisation and the redesign of the relationship between the State and local authorities have not brought about a great deal of increased power transferred to local authorities, even for the apparent discrepancies between the skills transferred and the financial opportunities. Rather, they have contributed to the rethinking of the general framework of centre/periphery relationships, with the introduction of partnerships between public and private actors, as well as of coordination and inter-institutional cooperation (Bobbio 2002). This is not a trivial result. Through legislative change, a process of redefining political and administrative action was induced: the introduction of flexible regulatory instruments has consolidated, and made formal and formalised, the interaction and the establishment of agreements between a number of actors and interests, facilitating the practical administration of such relationships. The centrality assumed by local authorities in a wide range of policies (from those relating to environmental issues to the one concerning the promotion of development) allowed to put in practice forms of vertical subsidiarity. Moreover, the positive acceptance, hence not as a binding constraint but rather as a possibility, of the plurality and the articulation of the actors and interests involved in urban and regional transformations has led to the opening up of the decision-making arena to traditionally excluded actors.

These elements can be traced down to the 'discovery' of the virtues of cooperation within a social dimension dominated by pluralism and the subsidiarity, and of competition as the main route to an efficient allocation of financial resources. Innovations introduced in the Italian context revolve therefore around two 'centres of gravity' within the rhetoric of current policy-making (Andersen and van Kempen 2003, Cochrane 2007, Gaudin 1999, Judge et al. 1995, Uitermark 2005): the first is the negotiation, associated with the constitution of agreements among central government, local authorities and private interests; the second is gradual shift of the role of public bodies, from a rather decision and regulatory role towards a role of *pilotage*, of direction or of 'accompaniment' of the interactions amongst actors (Jessop 1995, Kooiman 2003, Sibeon 2001).

According to Peters (2000), there is a significant difference between the traditional steering conception of governance, in which are still present forms of State coordination concerning the interactions between actors, policy priorities and contracts between the different actors and different interests (and is therefore conceived as a 'guide' to society and the economy), and the new modes of governance, whose distinguishing feature is both the plurality of interaction and the modalities of formal and informal regulation between public and private actors. It remains questionable into which of the two areas the Italian practices have entered with the greatest frequency and to the greatest extent. The amendment of Title V of the Constitution presents and expresses this ambivalence. This fits into the process of redefining the framework of powers between the State and local authorities (according to a traditional steering conception of governance). However, it also introduces a significant change in the mode of action of public bodies in urban government (which can be understood as an attempt to implement new modes of governance). The new Title V in fact exceeds the traditional scope of planning, a technical work closely related to the regulation of land use, which has traditionally operated urban planning in Italy, by introducing the reference to the term *governo del territorio*. This change in terminology in fact conceals a change in perspective and approach to the urban problems and to the procedures by which they could be governed, by 'breaking up', at least in words, the Italian practice of public intervention in the government of cities of a purely regulatory nature. Although the term *governo del territorio* is vague and ambiguous, 'una *buzzword* senza tradizione, che ammette una varietà poco ordinate di significati, che rinvano a pratiche ancora più diversificate' [a buzzword without tradition, with a variety of meanings which refer to more diverse practices], it can be understood as a 'extended field', indicating a passage from 'dalla sfera della semplice regolazione di usi e trasformazione del suolo a quello dello sviluppo insediativo e residenziale' [from the sphere of simple regulation of land use to the sphere of settlement and residential development] (Palermo 2009: 43). From this perspective, the *governo del territorio* is called to act through the spatial coordination of a variety of sectoral policies (from land use to landscaping, from mobility to the protection of ecosystems, from promotion of local development to the enhancement of cultural heritage), through the cooperation and coordination between multiple actors and interests as well as between different levels of government with the definition of partnership procedures, inter-institutional coordination and negotiated planning.

Criticism, however, is not lacking and some aspects do appear quite paradoxical. As emphasised by Gualandi (2007: 551), the spatial planning system that is emerging in Italy since the constitutional amendment, outlines the move from

> da un modello 'razionalista' e sistematico, nel quale alla gerarchia dei Piani corrispondeva una ben definite (e riconosciuta) gerarchia tra gli interessi ad un modello in cui se da un punto di vista teorico potrebbe apparire quasi 'eretico' ... parlare di gerarchia di interessi ... in realtà si assiste ... ad un fenomeno di accentramento e di concreta 'ascensione' dei processi decisionali, con l'individuazione di una pluralità di sedi e moduli decisionali, spesso alquanto estemporanei e privi di effettiva legittimazione. [A 'rationalist' and systematic model, in which to the hierarchy of plans corresponds a well-defined (and approved) hierarchy of interests to a model in which, even if from a theoretical point of view, may appear almost 'heretical' ... to refer to a hierarchy of interests ... in reality is possible to observe ... a phenomenon of centralisation and concrete 'ascension' of decision-making, with the specification of a plurality

of representation and decision models, often and somewhat improvised and lacking real legitimacy.]

In other words, the affirmation of the new keywords (subsidiarity, institutional pluralism, differentiation, citizen participation, etc.) is accompanied, in practice, with a centralisation of decisions which derives mainly from the need for speed and efficiency in decision-making, whatever the outcomes they may bring.

In the field of experimentation and practice, the picture is the same. Signs of change in the way Italian cities are governed are accompanied by persistent problems that continue to be ignored by the public agenda, that are treated poorly or in an entirely ornamental fashion, or even bypassed altogether by proposals that leave out the problems (and opportunities) of Italian cities (Dematteis 2011). These issues regard selected aspects of urban governance, for example environmental issues, in relation to energy saving in construction and building management or the reorganisation of traffic and mobility; quality of life and the increase of housing issues; the 'construction' of multicultural cities and the growing social exclusion of large segments of the population, etc. There are also more general issues related to the overall city government framework in Italy. One problem demonstrating the greatest inertia of Italian urban governance, which is often referred to but never acted upon, is the need to deal with the metropolitan government, both from the institutional point of view (with the non-imposition of metropolitan areas or cities) and, especially, in terms of substantial actions. To govern cities that are increasingly extended poses problems (and opportunities) of government that extend obviously beyond the boundaries of municipal competences. Italian urban governance, despite the experience of the strategic plans that 'question' the metropolitan theme, is however primarily a kind of local governance. This is linked to another perennial problem of urban governance in Italy, i.e. the traditional negligence of the state towards urban development strategies.

According to Donolo (2005), Italian urban policies have got, at least potentially, two types of innovations. The first type pertains to the policies based on interaction between public institutions and civil society, where the 'coinvolgimento di risorse della società … [serve] non solo per la copertura del consenso, ma anche per la formulazione e l'implementazione della politica' [involvement of society's resources … is there not only to attain consensus, but also for the formulation and implementation of policies] (Donolo 2005: 40). The second type refers to the policies focused on the interaction between public institutions and private business interests, whose distinctive feature is the 'costruzione di partnership fra attori portatori di interessi' [creation of partnerships between stakeholders] (Donolo 2005: 40). In the first case, the innovation lies in the possibility of producing public goods through social practices. In the second, 'l'innovazione maggiore … non sta nell'elemento negoziale, ma nei contesti regolativi in cui deve avvenire la contrattazione' [the major innovation … is not the feature of negotiation, but the regulatory context in which bargaining must take place] (Donolo 2005: 44) (that is: the presence of a regulatory framework imposing constraints, methods and evaluations). This distinction is reflected in those practices through which Italian cities are alternately (and uncritically) considered engines of economic growth, innovation centres and key actors to promote and strengthen international competitiveness, but also to allowing the development of several forms of self-organisation, which constitute, in reality, assistive devices to counter the shortcomings of the market (Jessop 2002). Therefore some schizophrenia emerges between an interpretation of cities as 'centres of competition', on the one hand, and as 'laboratories for new forms of social cohesion', on the other hand. These two interpretations of Italian cities correspond to the definition of policies separately aimed at economic development or at social well-being.

Furthermore, over the last 15 years, the concepts of competitiveness and cohesion have dominated the public agenda of urban policy not only in Italy (Ache et al. 2008, Boddy and Parkinson 2004, Buck et al. 2005, Fainstein 2001). Without actually considering the meanings of these two concepts, in the international debate and, more importantly here, in the Italian practices, the two terms are considered either simply in opposition or linked by a dependency relationship: cohesion is namely seen as a precondition for achieving competitiveness. In Italian cities, the theme (and the rhetoric) of competitiveness has been interwoven with the processes of structural transformation and growth of urban economies, with the growing importance of the hypothesis that cities function as 'collective actors' acting in a competitive context to obtain scarce resources (events, investments), with the real processes of physical and social changes, also related to the revitalisation of urban and real estate markets. Vice versa, the theme of social cohesion has been deployed in relation to the crises of the traditional modes of urban welfare, to the privatisation processes, and the outsourcing of public services, to the ongoing impacts of the two large urban demographic phenomena of the last decade (ageing and increasing immigration, legal and otherwise), to the new prominence of the housing problem and the emergence of security concerns. If, as highlighted by Fainstein (2001), the almost causal emphasis on the 'instrumental' relationship between competition and cohesion leads one to forget the value of cohesion in itself and is a rhetorical trick of a neo-liberal mould, recognition of the limits of a purely instrumental conception of the relationship between entrepreneurial characterisation of the city and social justice is becoming increasingly clear (Harvey 2008).

Overall, the change in Italian urban policies is reflected more in intention than in outcomes. The persistence of problems and the difficulties to enact changes in practices are two factors that characterise urban governance in Italy. Inertia, inefficiency and lack of attention to the problems of the city are combined with the progressive 'output stage' of cities as agents of public debate (cultural, social and political), except in terms of warnings (such as the urban safety-related to non-EU immigration, which has gained widespread coverage in Italian newspapers) and/ or major events (such as media campaigns related to the Winter Olympics in Turin in 2006 or to the next Expo in Milan). The final outcome of this intense period that appeared to point to renewal is entirely uncertain. Italy suffered substantial delay in the testing of more effective and efficient models of urban governance. As stated by Palermo (2009: 43–4), we remain poised between 'un rinnovamento ancora ampiamente incompiuto e una continuità sostanziale, che si limita solo a assumere forme discursive e manifestazioni empiriche apparentemente meno consuete' [a still largely incomplete renewal and a substantial continuity, which is limited to taking only discursive forms and apparently less usual empirical manifestations]. Despite the interest that it is potentially possible to find in some experimentations, when the lag is being made up, and action models already tested elsewhere are spreading in Italy, the Italian tendency is to produce and reproduce the most mundane and least rich in innovative potential experiences.

References

Ache, P., Andersen, H.T., Maloutas, T., Raco, M. and Tasan-Kok, T. eds. 2008. *Cities between Competitiveness and Cohesion. Discourses, Realities and Implementation*. Berlin: Springer.

Amato, F., Bolocan Goldstein, M., Cremaschi, M., Governa, F. and Pasqui, G. 2011. Torino, Milano, Roma, Napoli. Ciclo politico, agenda urbana, *policies* (1993–2010), in *Le Grandi Città Italiane. Società e Territori da Ricomporre*, edited by G. Dematteis. Venice: Marsilio, 207–60.

Amin, A. and Thrift, N. 2002. *Cities: Re-imagining the Urban*. Cambridge: Polity Press.

Andersen, H.T. and van Kempen, R. 2003. New trends in urban policies in Europe: evidence from the Netherlands and Denmark. *Cities*, 20(2), 77–86.

Atkinson, R. 2000. Combating social exclusion in Europe: the new urban policy challenge. *Urban Studies*, 37, 1037–55.

Bagnasco, A. and Le Galès, P. eds. 1997. *Villes en Europe*. Paris: La Découverte.

Balducci, A. 2000. Le nuove politiche della *governance* urbana. *Territorio*, 13, 7–15.

Belligni, S., Ravazzi, S. and Salerno R. 2008. L'élite che governa Torino. *Teoria Politica*, 1, 85–105.

Bevir, M. 2002. Una teoria decentrata della *governance*. *Stato e Mercato*, 66, 467–92.

Bobbio, L. 2002. *I Governi Locali nelle Democrazie Contemporanee*. Rome-Bari: Laterza.

Boddy, M. and Parkinson, M. eds. 2004. *City Matters: Competitiveness, Cohesion and Urban Governance*. Bristol: Policy Press.

Bolocan Goldstein, M. 2000. Un lessico per le politiche urbane e territoriali. *Territorio*, 13, 122–33.

Brenner, N. 1999. Globalisation as reterritorialisation: the re-scaling of urban governance in the European Union. *Urban Studies*, 36, 431–51.

Brenner, N. 2000. The urban question as a scale question: reflections on Henri Lefebvre urban theory and the politics of scale. *International Journal of Urban and Regional Research*, 24(2), 361–78.

Brenner, N. 2004. *New State Spaces: Urban Governance and the Rescaling of Statehood*. Oxford: Oxford University Press.

Buck, N., Gordon, I., Harding, A. and Turok, I. eds. 2005. *Changing Cities: Rethinking Urban Competitiveness, Cohesion and Governance*. London: Palgrave Macmillan.

Cammelli, M. ed. 2007. *Territorialità e Delocalizzazione nel Governo Locale*. Bologna: Il Mulino.

Cammelli, M. 2011. Governo delle città: profili istituzionali, in *Le Grandi Città Italiane. Società e Territori da Ricomporre*, edited by G. Dematteis. Venice: Marsilio, 335–78.

Carpenter, J. 2006. Addressing Europe's urban challenges: lessons from the EU URBAN Community Initiative. *Urban Studies*, 43, 2145–62.

Cassese, S. 2001. *La crisi dello Stato*. Rome-Bari: Laterza.

Chorianopoulos, I. 2002. Urban restructuring and governance: north-south differences in Europe and the EU URBAN initiative. *Urban Studies*, 39, 705–26.

Cochrane, A. 2007. *Understanding Urban Policy: A Critical Approach*. Oxford: Blackwell.

Conti, S. ed. 2002. *Torino nella Competizione Europea. Un Esercizio di* Benchmarking *Territoriale*. Turin: Rosenberg & Sellier.

Davoudi, S., Evans, N., Governa, F. and Santangelo, M. 2008. Territorial governance in the making. Approaches, methodologies, practices. *Boletín de la Asociación de Geógrafos Españoles*, 46, 33–52.

Debarbieux, B. and Vanier, M. eds. 2002. *Ces Territorialités qui se Dessinent*. Paris: L'Aube-Datar.

Dematteis, G. ed. 2011. *Le Grandi Città Italiane. Società e Territori da Ricomporre*. Venice: Marsilio.

Dente, B. 1999. *In un Diverso Stato*. Bologna: Il Mulino.

Donolo, C. 2005. Dalle politiche pubbliche alle pratiche sociali nella produzione di beni pubblici? Osservazioni su una nuova generazione di *policies*. *Stato e Mercato*, 73, 33–65.

Dowding, K. 2001. Explaining urban regimes. *International Journal of Urban and Regional Research*, 25(1), 7–19.

Fainstein, S. 2001. Competitiveness, cohesion, and governance: their implications for social justice. *International Journal of Urban and Regional Research*, 25(4), 884–8.

Faludi, A. and Janin Rivolin, U. eds. 2005. Southern perspectives on European spatial planning. *European Planning Studies*, 13, 195–331.

Gaudin, J.P. 1999. *Gouverner par Contrat. L'Action Publique en Question*. Paris: Presses des Sciences Po.

Geddes, M. 2000. Tackling social exclusion in the European Union? The limits to the new orthodoxy of local partnership. *International Journal of Urban and Regional Research*, 24, 782–800.

Governa, F. 2004. Modelli e azioni di *governance*. Innovazioni e inerzie al cambiamento. *Rivista Geografica Italiana*, 1, 1–27.

Governa, F. 2008. Teorie e pratiche di sviluppo locale. Riflessioni e prospettive a partire dall'esperienza italiana, in *Lo Sviluppo Locale al Nord e al Sud: un Confronto Internazionale*, edited by E. Dansero, P. Giaccaria and F. Governa. Milan: FrancoAngeli, 69–98.

Governa, F. and Saccomani, S. 2004. From urban renewal to local development. New conceptions and governance practices in the Italian peripheries. *Planning Theory and Practice*, 3, 328–48.

Governa, F., Janin Rivolin, U. and Santangelo, M. eds. 2009. *La Costruzione del Territorio Europeo. Sviluppo, Coesione, Governance*. Rome: Carocci.

Gualandi, F. 2007. Dal 'governo del territorio' al 'territorio … del Governo'?, in *Territorialità e Delocalizzazione nel Governo Locale*, edited by M. Cammelli. Bologna: Il Mulino, 551–68.

Gualini, E. 2004. *Multi-level Governance and Institutional Change. The Europeanization of Regional Policy in Italy*. Aldershot: Ashgate.

Harvey, D. 2008. The right to the city. *New Left Review*, 53, 23–40.

Hudson, R. 2005. Region and place: devolved regional government and regional economic success? *Progress in Human Geography*, 29(5), 618–25.

Imrie, R. and Raco, M. 1999. How new is the new local governance? Lessons from the United Kingdom. *Transactions of Institute of British Geographers*, 24, 45–63.

Janin Rivolin, U. 2003. Shaping European spatial planning. How Italy's experience can contribute. *Town Planning Review*, 74(1), 51–76.

Jessop, B. 1994. Post-fordism and the State, in *Post-fordism. A Reader*, edited by A. Amin. Oxford: Blackwell, 251–79.

Jessop, B. 1995. The regulation approach, governance and post-fordism: alternative perspectives on economic and political change. *Economy and Society*, 24(3), 307–33.

Jessop, B. 2002. Liberalism, neoliberalism, and urban governance: a state-theoretical perspective. *Antipode*, 34, 452–72.

Judge, D., Stoker, G. and Wolman, H. eds. 1995. *Theories of Urban Politics*. London: Sage.

Kooiman, J. 2003. *Governing as Governance*. London: Sage.

Lascoumes, P. and Le Galès, P. eds. 2004. *Gouverner par les Instruments*. Paris: Les Presses de Sciences Po.

Lauria, M. ed. 1996. *Reconstructing Urban Regime Theory*. Thousand Oaks: Sage.

Le Galès, P. 2002. *European Cities*. Oxford: Oxford University Press.

Logan, J.R. and Molotch, H. 1987. *Urban Fortunes: The Political Economy of Place*. Berkeley: University of California Press.

Mazza, L. 2000. Strategie e strategie spaziali. *Territorio*, 13, 16–28.

Mazza, L. 2011. Governo del territorio e pianificazione spaziale, in *Le Grandi Città Italiane. Società e Territori da Ricomporre*, edited by G. Dematteis. Venice: Marsilio, 261–316.

Molotch, H. 1976. The city as a growth machine. *The American Journal of Sociology*, 82, 309–31.

Mossberger, K. and Stoker, G. 2001. The evolution of urban regime theory. The challenge of conceptualization. *Urban Affairs Review*, 36(6), 810–35.

Osmont, A. 1998. La 'gouvernance': concept mou, politique ferme. *Les Annales de la Recherche Urbaine*, 80–1, 19–26.

Paba, G. 2003. *Movimenti Urbani, Pratiche di Costruzione Sociale della Città*. Milan: FrancoAngeli.

Palermo, P.C. 2009. *I Limiti del Possibile. Governo del Territorio e Qualità dello Sviluppo*. Rome: Donzelli.

Parkinson, M. 1998. *Combating Social Exclusions: Lessons from Area-based Programmes in Europe*. Bristol: The Policy Press.

Perulli, P. 2004. *Piani Strategici: Governare le Città Europee*. Milan: FrancoAngeli.

Peters, B.G. 2000. Governance and comparative politics, in *Debating Governance. Authority, Steering, and Democracy*, edited by J. Pierre. Oxford: Oxford University Press, 36–53.

Pierre, J. 2000a. Understanding governance, in *Debating Governance. Authority, Steering, and Democracy*, edited by J. Pierre. Oxford: Oxford University Press, 1–10.

Pierre, J. ed. 2000b. *Debating Governance. Authority, Steering, and Democracy*. Oxford: Oxford University Press.

Purcell, M. 2006. Urban democracy and the local trap. *Urban Studies*, 43(11), 1921–41.

Regonini, G. 2005. Paradossi della democrazia deliberativa. *Stato e Mercato*, 73, 3–31.

Rhodes, R.A.W. 1997. *Understanding Governance. Policy Networks, Governance, Reflexivity and Accountability*. Buckingham: Open University Press.

Rhodes, R.A.W. 2000. Governance and public administration, in *Debating Governance. Authority, Steering, and Democracy*, edited by J. Pierre. Oxford: Oxford University Press, 54–90.

Sibeon, R. 2001. Governance in Europe: concepts, themes and processes. *International Conference: Governance e Istituzioni: il ruolo dell'economia aperta nei contesti locali*. Forlì, Italy, 30–31 March.

Stoker, G. 1998. Governance as theory: five propositions. *International Social Science Journal*, 155, 17–28.

Stoker, G. and Mossberger, K. 1994. Urban regime theory in comparative perspective. *Environment and Planning C: Government and Policy*, 12(2), 195–212.

Stone, C.N. 2005. Looking back to look forward: reflections on urban regime analysis. *Urban Affair Review*, 40(3), 309–41.

Torino Internazionale. 2000. *Il Piano Strategico della Città*. Available at: www.torino-internazionale.org [accessed: 14 September 2011].

Torino Internazionale. 2006. *2° Piano Strategico dell'Area Metropolitana*. Available at: www.torino-internazionale.org [accessed: 14 September 2011].

Tosi, A. 1994. *Abitanti. Le Nuove Strategie dell'Azione Abitativa*. Bologna: Il Mulino.

Tubertini, C. 2007. Riforma costituzionale e sistema locale dei servizi sociali: tra territorializzazione e centralizzazione, in *Territorialità e delocalizzazione nel governo locale*, edited by M. Cammelli. Bologna: Il Mulino, 771–800.

Uitermark, J. 2005. The genesis and evolution of urban policy: a confrontation of regulationist and governmentality approaches. *Political Geography*, 24, 137–63.

Vandelli, L. 2000. *Il Governo Locale*. Bologna: Il Mulino.

Vanier, M. 1999. La recomposition territoriale. Un 'grand débat' ideal. *Espaces et Sociétés*, 96, 125–43.

Chapter 4
Urban Governance and the 'Profiles' of Southern Italy Cities

Carla Tedesco[1]

Introduction

This text discusses urban governance in Southern Italy focusing on the difficulties in giving actual meaning to innovative urban policies, especially to initiatives promoted and/or funded by the EU. The first assumption underlying this chapter is that, in order to discuss urban governance in Southern Italy, it is essential to overcome the traditional analytical framework of the large geographical division between the richer North-Centre of Italy and the poorer South (Mezzogiorno) – the so-called '*questione meridionale*' (literally 'southern question', the social, economic and cultural backwardeness of Southern Italy) has been drawn on since the late nineteenth century. Southern Italy cities can be described in several different ways, for instance as Mediterranean cities, not only as cities located in a 'lagging' region. On the other hand, discussing urban governance can be considered to be a way to overcome the idea of the Italian Mezzogiorno as a whole. This idea has been increasingly abandoned by academics and politicians during the late twentieth century in favour of a plural vision which stemmed from historic, social and economic studies. These will be explored in the following section.

What is important to underline here is that the new visions concerning regional development in Southern Italy which were produced by these studies can be related to other research into the role of territorial governance as an intermediate level of political, economic and social regulation. During the 1990s this role emerged as worthy due to many processes, in particular: the gradual weakening (or reorganisation) of the Nation-State's competences in the economic domain and the new relationships between the Nation-State political domain and the globalisation processes; the fragmentation of national societies and their traditional institutions (political parties, trade unions, churches); the EU integration processes (Le Galès 1998, Le Galès and Lequesne 1998). As a consequence, during the 1990s the renewed interest for regions, which had already emerged since the 1970s, was reinforced (Sabel 1988) and cities started to be considered more and more

1 IUAV University of Venice; Department of Design and Planning in Complex Environments. Ca' Tron; Santa Croce 1957; 30135 Venice (Italy). ctedesco@iuav.it.

as new units of analysis (Pichierri 1999). In particular, European cities were acknowledged to be places for cultural exchanges, social relationships, invention of new government techniques (Le Galès 2006: 16). Thus, the importance of urban political actors increasingly emerged throughout Europe and urban governance became central to the scientific and political debate.

Governance has been defined in several different ways. It is not the aim of this chapter to discuss them. We refer to governance 'to indicate a new mode of governing that is distinct from the hierarchical control model, a more cooperative mode where state and non-state actors participate in mixed public/private networks' (Mayntz 1998). According to Le Galès (2006: 42), we assume governance as a notion useful to raise questions rather than to give answers. And this is the second assumption underlying this chapter.

These ideas on urban and territorial governance influenced policy-making in various European countries. Many new public policies were set up and implemented; these assumed a less hierarchical form, overcame the traditional policy sector boundaries and were supposed to be less influenced by powerful interest groups (Lascoumes and Le Galès 2004: 23). Territorially-based initiatives assuming the form of 'social pacts' multiplied during the 1990s through a process of institutionalisation the European Commission (EC) had a decisive influence in promoting and generalising (Pichierri 2002). A number of types of 'local concertation' initiatives were emphasised, among them: public–public partnerships, forms of 'industrial adjustment concertation', second generation industrial districts, strategic urban planning, projects funded by European programmes, territorial pacts (Pichierri 2002). They were established across different types of target areas and assumed different focuses including initiatives aiming at: tackling multiple deprivation problems in urban neighbourhoods; promoting integrated and participative approaches to rural development; fostering local economic development and employment in subregional areas etc. Many of these initiatives were promoted and/or funded by the EU Structural Funds policy.

As far as urban policies are concerned, as is well known, during the 1990s and 2000s the EU, namely the Commission of the European Communities (CEC), increasingly drew attention towards cities and urban issues due to the importance attached to the subsidiarity principle by the Maastricht Treaty. It stressed the urban dimension of EU and national sectoral policy and promoted and/or funded area-based urban regeneration programmes (Atkinson 2001, 2007, Cremaschi 2005, Parkinson 2005). Furthermore, the EU directly supported, within Structural Funds policy, the formation of transnational urban networks, including interconnections and transfer mechanisms between cities (Atkinson and Rossignolo 2010). According to Lascoumes and Le Galès (2004: 25): 'La création d'instruments d'action publique peut servir de révélateur de transformations plus profondes de l'action publique, de son sens, de son cadre cognitif et normatif et des résultats' ['The setting up of policy tools can reveal deeper transformations of public action, its sense and cognitive and normative framework as well as of its outcomes']. This brings us to our third assumption: the capacity to set up (and to implement) innovative urban policies is a peculiar aspect of urban governance which is worth exploring.

Within this framework, we focus on a very specific issue: the difficulties in giving actual meanings to new policy instruments during their setting up and implementation. This seems to be a common feature of urban policy processes in Southern Europe: 'It would therefore seem that for now (with some notably and praiseworthy exceptions), although new means of urban governance in Southern Europe show great potential in the realm of theory, these continue to present serious difficulties in terms of their actual implementation' (Seixas and Albet 2010: 781). This is true also for Italy (Governa 2010: 674) and especially for Southern Italy (Tedesco 2006, 2009). Investigating the reasons for these difficulties is the main point addressed in this chapter.

We focus, in particular, on urban policy initiatives either directly promoted and/or funded by the Commission of the European Communities (CEC) or promoted and funded by Structural Funds Regional Operational Programs. The former have been acknowledged to play an important role within the evolution of urban policy in Italy during the 1990s and 2000s from a top-down and sector oriented approach towards a bottom-up and participative approach. We will briefly describe this shift in the third section (Balducci 2000, Cremaschi 2003, Palermo 2001). The latter help us in highlighting local interpretations of EU policy frames. These local interpretations are worth taking into account, as they played a very significant role in the mainstreaming of the EU approach to urban issues into Structural Funds Regional Operational programmes. Although we focus on urban initiatives promoted and/or funded by the EU, national policies undertaken in the same years will be briefly mentioned in order to better understand the general framework of urban policy in Italy during the 1990s and 2000s. Furthermore, although, in a governance perspective, actors, groups, interests have an importance even beyond local administrations, we take particularly into account the latter as a key element to understand the capacity to set up and implement innovative effective policies.

How can we analyse the interplay between cities and the EU? In recent times scholars have increasingly endeavoured to understand it by the concept of Europeanisation as, broadly speaking, 'the impact of the European integration process on the EU member states (but also on the EU accession countries) [which] has become a key concept in explaining political and societal change across the continent' (Hamedinger and Wolffhardt 2010: 10). In particular, Marshall (2004), drawing on evidence from British cities, has developed a framework for evaluating Europeanisation at the municipal level. He distinguishes two different mechanisms: 'download Europeanisation', where the negotiation and implementation of EU programmes results in changes in local policy and practice; and 'upload Europeanisation', where innovative local initiatives are incorporated in pan-European policies or programmes (Marshall 2004). Previously, referring to Marshall's framework and focusing on two Italian cities participating in an EU programme supporting the development of urban networks (the URBACT network),[2] I highlighted that Europeanisation processes were twofold, involving both download and upload mechanisms. I argued that interconnections and transfer mechanisms between cities within urban networks do not necessarily have an influence on EU policy-making, especially in Southern Italian cities. Instead, they are likely to produce impacts on local policy and practices (Tedesco 2010).

In this chapter a quite different perspective is assumed. The different 'attitudes'of Southern Italian cities towards implementing EU (or EU 'style') initiatives are considered to be a track of their distinctiveness. What if we try to gather Southern Italian cities relating them either to some specific problems they possess or to some specific local resources they can count on rather than assuming their location in a (relatively) underdeveloped area as the main frame to describe them? What problems do they highlight and what resources do they use in order to participate in EU competitive bidding processes?

To sum up, given the fragmented landscape of Italian 'Mezzogiorni', as it emerged mainly from historical and socio-economic analysis during the late twentieth century, this chapter discusses urban governance in Southern Italy. It focuses on the modes cities endeavoured to give actual meaning to EU innovative urban policies. Even if we assume a specific focus for our discussion, this is not at all an easy task. Hence, this chapter can be considered to be an 'assignment in progress'. It discusses the points for a research agenda more than illustrating research results.

This chapter is divided into three sections following this introduction. In the first section we start with a brief review of studies which contributed to abandon the framework of the Italian Mezzogiorno as a whole and to consider this area both within the EU and the Mediterranean. In the second section, after a brief description of EU innovative urban policies set up and implemented in the last 20 years, some evaluations of their outcomes in Southern Italy are discussed, focusing, in particular, on the relationships between policy innovation and administrative innovation. In the third section the possibility to relate these outcomes to the specific profiles of Southern Italian cities by investigating their participation in EU initiatives and networks is analysed referring to a case-study about Apulia region cities. In the fourth and last section some issues are discussed concerning multi-level governance and the coexistence of 'old' and 'new' in Southern Italian cities.

2 The EU URBACT programme is devoted to 'develop transnational exchanges of experiences between actors, whether cities or other partners' included in the EU urban programmes (CEC 2002). 'Good practice transfer' is a central element of the URBACT programme. URBACT has continued since 2007 under the name of URBACT II. This manages the exchange of experience among European cities and the capitalisation and dissemination of knowledge on all issues related to sustainable urban development. The objectives of URBACT II are not only promoting the exchange of experience and the diffusion of good practice; but also assisting city policy-makers to define local action plans for sustainable urban development.

Framing Southern Italian Cities between 'Mezzogiorni', the EU and the Mediterranean

During the late twentieth century, the idea of the Italian Mezzogiorno as a whole has been increasingly contrasted from several different perspectives. Some historians shed light on the different reactions of 'Mezzogiorni'[3] to the processes of modernity[4] by both taking into account the different centre–periphery relationships within the Mezzogiorno and analysing the different interdependencies between multiple internal and external places.

From a different perspective, some sociological studies had developed a new geography of the Italian production system, overcoming the traditional North–South divide by the acknowledgement of a 'Third Italy'. In Northern Italy two different development models had been recognised: in the North West it was based on big enterprises, in the North East it was based on small and medium enterprises. Each of them had specific features in terms of society, political system, culture (Bagnasco 1977). Although within this new frame the Mezzogiorno was still considered as a whole, this study paved the way to the acknowledgement of modes of production different from the big enterprise.

In the following years, other socio-economic studies (see, in particular, Bodo and Viesti 1997, Cersosimo and Donzelli 2000, Meldolesi 1998, Viesti 2003) recognised in the Mezzogiorno many interesting tracks of endogenous development, small and medium enterprises (SMEs), forms of industrial districts, local clusters well integrated in the national and international markets, technological poles. As a consequence, the traditional image of the Mezzogiorno as an homogenous underdeveloped area was not valid unymore. Furthermore, the unintended consequences of public policy for Southern Italy were highlighted (Trigilia 1992). This had produced at the same time modernisation and dependence: the per capita incomes had increased, however this had not involved endogenous development taking off. These studies pointed out that public policy itself, as it had been conceived after the Second World War, could be considered to be an obstacle for local economic development.

Thus, discussions about the conditions for accelerating economic development by fostering local clusters of economic activity developed (Viesti 1999). This would have brought about a 'big shift' (Bodo and Viesti 1997). To put it briefly, a new idea of development stemmed from this debate (Donolo 1999). This was to foster local, bottom-up development abandoning the top-down approach. This

3 In the Italian language, when forming the plural of nouns, the endings change to indicate a change in number. For regular masculine nouns, which end in -o in the singular, the ending changes to -i in the plural.

4 We refer, in particular, to the studies published since 1987 in the journal *Meridiana* by IMES (Istituto Meridionale di Storia e Scienze sociali, Southern Institute for Historic and Social Studies) which gathered contributions by historians and other social scientists.

idea was supported by EU policies following the 1988 Structural Funds reform (Tofarides 2003).

Another influential framework developed in the late twentieth century, which contributed to overcome the traditional dualistic vision of the Italian territory, was centred on the Mezzogiorno's role within the Mediterranean. A 'Mediterranean way' for development drawn on the role Southern Italy could play within the Mediterranean as a 'junction' between North and South, as well as between East and West, was specified (Cassano 1997, quoted in Masella 2009). These ideas involved both a federalist perspective and the rise of the civil society free from the party system. They can be somehow linked to investigations into the Mediterranean city which concentrate on postmodern cultures evident in everyday urban life and recognised elements of postmodernism in the social and cultural life of Mediterranean cities (Leontidou 1993),

Hence, on the one hand, the debate about 'Mezzogiorni' and, on the other hand, both the EU and the Mediterranean horizons contributed to overcome the traditional dualistic perspective of the Mezzogiorno as an homogenous underdeveloped area in contrast with the richer North of Italy. This opened new perspectives. Following the above considerations we state that as far as Southern Italian cities are concerned, being located in a (relatively) underdeveloped area is not the only way they can be described. We will look for other descriptions analysing the modes in which they implemented EU urban policies which stemmed from the above debate. Before continuing a word about EU urban policy instruments is necessary.

Urban Policies in Southern Italian Cities

In Italy there is not an explicit national urban policy, but, during the 1990s and early 2000s a 'constellation' of initiatives were undertaken by different (local and central) public institutions (the EU, the national government and some regional governments); these aimed to introduce either procedural innovations or specific programmes of action both area-based (in the field of urban regeneration, social issues and poverty or local economic development) and sector oriented (in the field of unemployment and training, education, criminality) (Padovani 2010: 38). A comprehensive review of these new instruments will not be given in this chapter (cf. Briata et al. 2009, Governa and Saccomanni 2004, Palermo 2001). However, we need to underline that, in order to deal with these innovations, public administration was required 'to shift from the more consolidated logic of providing subsidies, incentives, or constraints toward a new logic of action more open to cooperation, negotiation, and dialogue between the different governmental departments, between local and central government, and between public, private and volunteer sectors' (Padovani 2010: 38).

In particular, all these new policy instruments (the EU and the national ones) have introduced significant changes in the Italian urban governance systems: the explicit acknowledgement of the presence of new actors involved in urban changes; the opening of the decision-making process to new modes of collaboration and partnership; the growing importance assigned to both policies aiming at putting the city into the international competition circuit and policies to tackle social exclusion in urban deprived areas (Dematteis et al. 1999). However, two of the groups alone, which are particularly significant for this discussion, will be considered: area-based urban regeneration programmes directly promoted and funded by the Commission of the European Communities (CEC) (in particular, URBAN Pilot Projects-UPP, launched in 1990; the URBAN and URBAN II programmes, launched in 1994 and 2000); and urban policies funded by the Structural Funds Regional Operational Programmes (ROPs), during the 2000–6 and 2007–13 programming periods.

Nowadays a general disappointment emerges from several evaluations of the implementation of urban and regional policy in Italy in the last 15–20 years (Alulli 2010, Barca 2006, Bricocoli and Savoldi 2010, Cremaschi 2010, Laino 2010, Pasqui 2010, Viesti 2009). On the other hand, some authors underline the need to overcome 'localism' in favour of a top-down approach assuming that the North–South divide is still persistent and aiming at grasping the opportunities related to the central position of the Mezzogiorno in the Mediterranean. Policies aiming at improving the efficiency of local institutions as well as building up big strategic infrastructures, that Southern Italy still requires, are the central elements of this view (Rossi, quoted in Masella 2009: 485–6).

Certainly, although in the last 20 years many innovative policies have been promoted and funded, the economic gap between Northern and Southern Italy is increasing. The economic trends of the South are always worse than those of the North (Viesti 2009: 137). However, to recognise the persistence of a gap at the regional level does not necessarily mean to forget subregional differences. Some authors have shed light on the different reactions of both Northern and Southern Italian regions to state intervention aiming to reduce regional socio-economic disparities since the late nineteenth century until recent Structural Funds policy, taking into account not only the per capita income, but also cultural factors (Felice 2007). The reactions of cities, instead, have not been deeply analysed in a comparative perspective although cities have been increasingly recognised in most EU documents as the 'engine' for regional development (CEC 1997, 1998, 2006). At the moment, an evaluation of urban policy outcomes seem to be very difficult. A lot of research is needed before we fully understand them. Nevertheless, we can now give some elements for interpretation.

The above-mentioned evaluations seem to have either forgotten or left in the background the specific features of Southern Italian local contexts which had emerged from the above-mentioned studies criticising the traditional top-down state intervention, which had been the premise for the shift towards a bottom-up approach to development policies. Taking into account all these specific features is not possible in a short chapter like this. Only one of these issues, which, in our opinion, is particularly significant for this discussion, will be considered: the difficult relationship between policy innovations and administrative innovations.

On the one hand, given the traditional 'weakness' of both Southern Italian local administration institutions and urban civil society (Trigilia 1992: 88), the setting up and implementation of the new bottom-up, area-based and integrated policies (which had multiple objectives and included several different kinds of actions and actors) was a very ambitious objective: it is not surprising at all that these policies involved many difficulties during the policy process, including frequent schedule delays and disappointing outcomes. Besides, assessing Southern Italian local administration performances in dealing with innovation without taking into account local organisational structures as they result from historic clientelistic processes is perceived as very unfair by new actors endeavouring to introduce innovation (Cipriano 2010).

On the other hand, if one focuses on the relationships, at the 'micro' level, between urban policies' local outcomes and the modes of organisational innovation in terms of learning within public administration taking into account not only policy changes, but also work practices changes (in recent decades there has been a dramatic increase in precarious work), it is possible to highlight that what is learnt somehow does not automatically become part of the know-how of local administrations, i.e. often innovative practices experienced by a distinct department of a distinct public institution within the implementation of a specific programme have been transferred either into other departments of the same institution or into other public/private organisations. And all the more so, as integrated urban policy involves several dimensions and objectives, thus people involved in them acquire skills which can be used in several policy sectors (Tedesco 2011).

What is more, the problems of Southern Italian cities, such as, more generally speaking, the problems of Southern European cities, do not correspond exactly to the ways in which cities were defined as a policy problem by EU documents (Chorianopoulos 2010). Hence, local administration institutions have to both interpret EU policy frames and understand how to match local needs (which often largely differ from the needs of Central and Northern European cities). For instance, the acknowledgement of pockets of deprivation within wealthy cities and regions was the basis EU urban policies were drawn on. However, this does not fit the situation of pockets of severe deprivation within 'lagging' cities and regions which are often characterised by informal settlements as well as by informal economy.

> The lack of State control and large developers in the housing construction sector, combined with land fragmentation, result in piecemeal urban development and a large informal housing sector … This overall dispersal and intermingling of land use is not only due to anti-planning attitudes and the fragmentation of property ownership: it is also due to the widespread informal economy during a long period going back in history. (Leontidou 1993: 953–4)

Thus, if breaking the barriers between pockets of deprivation and the rest of the city is a way to foster new development paths in Northern and Central European deprived areas, this is a very meaningful objective, but very often not at all sufficient to foster local development in Southern Italy. Besides, self-built, often either unauthorised or retrospectively legalised, peripheral settlements coexist in Southern Italy cities with planned large social housing neighbourhoods as well as with the formal and informal (often illegal) economy.

We argued elsewhere (Tedesco 2005, 2006) that within the implementation of EU programmes 'hybridisation' processes between the EU policy frames and the local ones occurred. On the one hand, these hybridisation processes can be considered to be a way to match local needs and European opportunities; on the other hand, they underline the development of a capacity to benefit from funding opportunities, offered by Structural Funds. Thus, these policy initiatives, at the same time, were creating or contributing to create development and were a different way of being dependent on the 'top' (Tedesco 2003). Nowadays, we can state that some local contexts seem to have undertaken new development paths, opening the urban arena to new forms of cooperation, negotiation and dialogue between public, private and voluntary sectors.

In order to investigate the features of these processes, in the following section we refer to cities located in a distinct Southern Italy region, Apulia, and we analyse their involvement in EU Structural Funds policies.

A Case-Study

The Apulia region (Regione Puglia) is located in the South East of Italy, along the Adriatic coast. It used to be divided into five provinces. Each of them has a main town: Brindisi, with about 90,000 inhabitants, is a port town on the Adriatic coast; Bari, with about 330,000 inhabitants, is the regional main town and also a port town; Foggia, with about 150,000 inhabitants, is located in the middle of an important agricultural area; Lecce, with about 95,000 inhabitants, is located in the very south of the region, in an area which has been marginal until to the 1990s, as it was excluded from 1950s and 1960s industrial development national strategies, and has been experiencing touristic development since the 1980s; Taranto, with about 200,000 inhabitants, is a port town on the Ionian coast whose economy used to be based on the steel industry (one of the largest steelworks in Europe was established in Taranto in the early 1960s, it was privatised in 1996), it is nowadays experiencing severe unemployment and environmental problems. A sixth province was established in 2004. The latter is has got three main towns: Barletta (about 90,000 inhabitants), Andria (about 90,000 inhabitants) and Trani (about 50,000 inhabitants). These towns share a manufacturing cluster.

As far as the regional structure is concerned (cf. Borri 1996), the development policies of the 1950s and 1960s were drawn on the dual vision of coastal (developed) areas/internal (underdeveloped) areas. Industrial poles were established in Bari, Taranto and Brindisi. These visions were confirmed during the 1970s. Another influential analytical framework highlighted the polycentric character of Apulia region's urban structure. However, if one looks thoroughly, this is not the case in the Northern and Southern part of the region, characterised by very small towns.

During the early 1990s, in the Apulia region innovation in urban policy was mainly linked to the availability of some new programmes either launched by the EU or by the national government. As far as EU programmes are concerned, during the 1994–9 Structural Funds programming period, an URBAN Pilot Project (II round) was undertaken and implemented in Brindisi and three URBAN programmes were carried out and implemented in Bari, Foggia and Lecce. In all three cases, URBAN was mainly perceived as an innovative and successful experience. This perception was linked to some distinct aspects of the programme: the capacity of the local administration to deliver the programme following the procedures and strict deadlines of EU Structural Funds, and the capacity to coordinate different sectors within the local administration.

This success was limited to a few experiences (and the people involved in them). It was particularly evident as the social housing area-based programmes launched by the national government in the same period were very slowly implemented and were often perceived as failures.

Afterwards, within the 2000–6 programming phase, two URBAN II programmes were undertaken and completed in Taranto and in a small town located in the Bari metropolitan area (Mola di Bari, about 25,000 inhabitants). In the same period the 'URBAN approach' was somehow mainstreamed: an area-based programme targeting deprived neighbourhoods within the mainstream of Structural Funds Regional Operational Programs for the 2000–6 programming period was set up and launched by the regional government; this was included in the 'Cities priority', funded by three different measures (devoted to physical, social and economic actions[5]) and included in the ROPs, whose funds were partly (or wholly) assigned to these programmes and targeted at the former five provincial main towns. Cities interpreted this programme as a 'mainstreamed' URBAN. But it was perceived as a failure in comparison with the URBAN experience. Furthermore, the stress on the 'physical' aspects of regeneration (interventions on the built environment and the environment) which is typical of the Italian interpretation of the URBAN programme, largely prevailed.

Hence, one can say that in the mid 2000s, although a few innovative area-based programmes had been experienced, urban policy arenas had not been opened to the new bottom-up and integrated approach. However, in this period, some cities, in particular Bari, Lecce and Taranto, started to set up proposals for urban strategic plans which were supposed to involve significant change in urban governance.

In the same years (namely in 2005), a new left-wing regional government, willing to break strongly with the past right-wing government policy approach, was elected. The election of the new regional government coincided with a 'season' of further innovations in dealing with urban issues. An academic expert in the field

5 In particular, actions on the built environment and the environment were funded by the cities priority, social and economic actions by other priorities.

of urban policy and planning was appointed as a planning, urban policy and social housing councillor. The same regional government was re-elected in 2010, hence, innovative urban programmes started to be strongly promoted also by the regional government. Two parallel processes started in the same period: the introductions of deep innovations in the spatial planning system and in urban policy and the setting up of the 2007–13 EU regional policy programming period documents.

Within this framework, we will focus on two distinct processes, which are particularly significant for our discussion: the launch of urban regeneration initiatives by the Planning, Urban Policy and Social Housing Department and the setting up of subregional strategic plans, promoted and funded by the the Budget and Programming Department (the Managing Authority of the ROP). We will analyse these processes mentioning, in particular, some cases where the processes carried out are particularly indicative of differences between cities.

As far as urban regeneration initiatives are concerned, an area-based and integrated programme aiming at grasping the specific problems of urban deprived neighbourhoods, including social housing to be renovated or built, was set up and launched by the regional government in 2006. This was the Programma Integrato di Riqualificazione delle Periferie (PIRP, Peripheral Neighbourhood Rehabilitation Integrated Programme) (cf. Tedesco 2009). The innovations which the programme introduced in the local contexts concerned formal procedures, programme content and participation practices. It strongly stressed integration (between housing, infrastructure, social and economic actions), sustainable building and the involvement of local people in the regeneration process. The three key words of the programme were: integration, sustainability and participation.

After the launch of the programme, the attempt to introduce innovation in terms of procedures, content and social practices was supported through the organisation of workshops which aimed at diffusing concepts such as integration and participation; it was also supported through the establishment of an online forum supporting municipalities in the setting up of programme documents. The supporting initiatives were carried out in cooperation with a new department set up at the regional level, the 'Transparency and Active Citizenship Department', in charge of setting up participatory processes in all the policy fields.

The PIRP was successful. Many municipalities participated in the competitive bidding process: 123 out of 258 municipalities in the Apulia region presented 129 programmes (the province's main towns could present two programmes). What were the reasons for this success? Worth mentioning here is that the funds available were not particularly consistent,[6] thus, other factors urged municipalities to participate. One of these was certainly the possibility of getting the programmes (and the planning permissions) approved (even if not funded) by the regional administration in quite a short time. But it is not the only one. A cultural shift towards new approaches to urban issues was occurring (Tedesco 2009).

6 Each local programme could benefit from €4m (for cities with a population of at least 50,000), €3m (for cities with a population of at least 20,000) or €2m in other cases.

The analysis of the implementation of this programme clearly highlights that practices of 'resistance to change', intentionally and unintentionally developed by some of the different actors involved in the policy process, at both regional and local levels emerged: changes in administrative and governance structures are much slower than changes in policy instruments. Traditional organisational weakness is a main check on innovation in Southern Italy even when a political change pushes towards innovation. However, the regional government carried on with promoting innovative policies. In the following years area based and integrated urban regeneration initiatives based on public/private partnerships and on the involvement of local people were mainstreamed by a regional law (l.r. 21/2008).

Moreover, in the ERDF ROP 2007–13 a distinct priority devoted to cities was established. This was Priority VII, 'Competitiveness and attractiveness of cities and urban systems'. Within this priority two kind of actions were established. The first one targeted networks of small towns: the second one targeted bigger towns.[7] The approach carried out by the regional government produced new forms of territorial governance: new forms of horizontal and vertical cooperation between institutions have been eventually experienced as well as innovative modes of citizens inclusion into regeneration processes (Pace forthcoming). At present (January 2012) the most meaningful outcomes in terms of production of new forms of knowledge for sustainable urban development through new modes of cooperation concern small towns; in particular, small towns located in marginal areas (in the Northern and Southern part of the region) excluded not only from industrialisation processes, but also from more recent processes of touristic development (Pace forthcoming). In these towns, regeneration programmes have drawn on urban practices linked to both everyday life and cultural events opening local contexts to national (and in some cases international) networks.

Even referring to bigger towns, it is in Lecce, excluded from industrialisation poles strategies in the 1950s and 1960s, where at present interesting changes in urban governance are more evident (Parlangeli forthcoming). In fact, urban regeneration initiatives have been framed within a subregional strategic plan aiming, on the one hand, at developing institutional capacities and, on the other hand, at fostering integrated and participative urban policies. Before continuing a word about subregional strategic plans is necessary.

Subregional strategic plans spread since the mid 2000s in Southern Italy as they were funded within the national regional policy, which in Italy for the first time during the 2007–13 programming period was unified with EU Structural Funds policy in terms of time schedule and definition of objectives, priorities and rules. In Regione Puglia they were activated following a competitive bidding process launched by regional governments in 2006. Ten strategic plans were

7 The information in part of this chapter has been acquired from documents, interviews and informal discussions with privileged witnesses at the regional government level and at the municipal government level. The author of this chapter has also been involved in the setting up of Priority VII of the 2007–13 ERDF Programme Document.

funded. Following negotiations between the regional government and the local governments, all the Apulia region municipalities were included in these strategic plans, which were related to the ERDF Operational Programme implementation.

In 2007–13 cohesion policy documentation there is an emphasis on the role of the 'territorial dimension':

> The renewed Cohesion policy for 2007–2013 encompasses a more strategic approach aiming to boost and integrate growth strategies at European, regional and local level, taking into account of the territorial dimension and specificities of regions and based on a reinforced partnership. (CEC 2006: 2–3)

Thus, 'subregional strategic plans' became the main policy instrument to foster the territorial dimension of Structural Funds policy as well as an instrument to experience new modes of governance.

According to Mazza (2004: 127) we refer to 'strategic planning' as:

> a voluntary political-technical action aiming at building up a coalition sharing some strategic lines (the 'strategy'); the coalition willing and being able to implement the strategy singled out. As a consequence, a strategic action is essentially informal, strategic planning cannot be prescribed by law, as building up a coalition cannot be prescribed by law.

Notwithstanding the different meanings and experiences of strategic planning in the Italian context we can assume as distinctive features of these processes the setting up of a consensus-based spatial vision and the voluntary character of mobilisation of public and private actors around some shared actions (cf, among others, Curti and Gibelli 1996, Fedeli and Gastaldi 2004, Martinelli 2005).

Referring to the voluntary character, which we pointed out above as a main feature of strategic plans in Italy, the 'odd' nature of the processes which took place in the Apulia region is evident: they were activated due to funding opportunities for both the design of the planning documents and the implementation of some actions included in it. This, in most cases, without any awareness about what strategic planning is and what its potential for territorial development is.

It is very difficult to look at these processes as a whole. Although in some cases subregional strategic processes opened the urban arenas to new actors, it is difficult to affirm that they either fostered or somehow contributed to an evolution from government to governance. Government and governance emerged rather as coexistent and, for many aspects, separated 'worlds'. However, while a few years ago these worlds differed strongly in terms of actors included and expertise developed, nowadays in some cases a crossover seems to emerge. This is the case of Lecce, which we already mentioned. By contrast in the case of Taranto, for instance, many severe problems are still at stake. Since the early 2000s local government sought to replace both negative trends of decay and the image of the city as a polluted 'steel town' by the setting up of a strategic plan. The most vivid image produced within this process concerned Taranto as a port town. However, these images seemed to divert attention away from other existing and possible images of the town, mainly produced by Structural Funds policies, also due to the scarce involvement of local stakeholders and citizens in strategy making (Barbanente and Monno 2004). The more recent participation of Taranto in the subregional strategic planning process fostered by the regional government still raises doubts concerning Taranto as a typical example of a city considered unable to change (Monno 2010).

Some Open Issues

Our analysis of the processes of implementation of EU Structural Funds programmes in Apulia region cities highlights that also when located in the same region, Southern Italian cities cannot be at all considered to be homogenous entities, in spite of their location in the same (relatively) underdeveloped region. Further in-depth comparative research is needed to fully understand these processes. However, in our opinion, two main features of Southern Italian cities clearly emerge from our work. In this section we will relate them to urban governance issues.

First, the coexistence of 'old' and 'new'. Leontidou (1993: 961) highlights the 'complex interweaving of traditional modern and postmodern conditions which in fact gives Mediterranean cities their complex palimpsest'. The effectiveness of this new urban policy, which fosters new forms of both horizontal and vertical cooperation, negotiation and dialogue between public, private and voluntary sectors, is strictly linked to the way 'old' hierarchical relationships and 'new' cooperative relationships between local actors are mixed up in the cities.

Referring to our case-study, at present, although the size of the cities seem to be an influential factor in respect to its influence on the capacity for setting up and implementing effective policies, with smaller towns being more able to benefit from EU Structural Funds opportunities than bigger towns, if one looks at it more thoroughly, it is the fact of having either been excluded or included in development policies based on the modernist paradigm which emerges as a crucial element. Cities like Lecce, which were not at all interested by the 1950s and 1960s 'growth poles' industrial development strategy seem to be more easily benefiting from new urban policies and strategies. By contrast, new urban policies seem particularly difficult to apply to situations in which the paradigm of modernity cannot be so easily dismissed. Such is the case of Taranto, the 'steel city'. In July 2012, after a long inquiry into whether dioxin and other chemicals pumped from the plant caused abnormal rates of cancer and respiratory and cardiovascular diseases, a partial closure order has been decided and eight executives have been placed under house arrest. As a consequence about 5,000 people have joined a strike to protest the threat to (about 11,000) jobs.

The end of Fordism and the actual postmodern condition seems to produce as a main outcome in the Mezzogiorno a deconstruction of traditional social networks.

The centrality of the service sector, which is typical of the postmodern era, goes hand in hand with the spreading of traditional economic activities (both legal and illegal) and with the survival of fragments of the industrial system. In areas with high unemployment rates, such as the decaying industrial poles in 'lagging' regions, environmental externalities are often accepted as a price which is worth paying in order to avoid the threat to jobs (Masella 2009: 489). This difficult balance between environmental externalities and jobs is crucial. These 'remnants' of modernist development strategies act as a check on new development strategies. Their influence is certainly stronger than the traditional institutional weakness.

Kunzmann (2010) describes the medium sized towns which are forced by globalisation processes to find their own profiles between international orientation and local embededness. In his view towns like Lecce, located in the periphery of Europe, can be considered to be among the relative losers of globalisation. What emerges from our analysis somewhat challenges Kunzmann's view and shows us possible directions for further research which is certainly required.

Second, coming back to our Europeanisation framework, download Europeanisation processes in Apulian cities seems to be particularly interesting as they have been reinforced by the regional level. What is more, in this case, it is particularly evident that it is not possible to evaluate change in urban governance if one does not take into account the regional (and national) levels. Urban governance needs to be considered from a multi-level perspective.

Last, but not least, as we underlined above, the acknowledgement of pockets of deprivation within wealthy cities and regions was the basis EU urban policies were drawn on. However, this description did not fit the situation of pockets of severe deprivation in Southern Italy, where we are often ahead of deprived urban areas in poor regions. These problems are likely to increase in times of economic crisis. At the same time, economic crisis is likely to spread somehow this kind of situation. An increasing number of EU cities located in wealthy regions are suffering from socioeconomic problems due to the global economic crisis. Hence, both for them and for the cities located in 'lagging regions', urban deprivation problems are not only those of breaking barriers between the existing jobs and the people who are excluded from the labour market.

References

Alulli, M. 2010. *Le Politiche Urbane in Italia. Tra Adattamento e Frammentazione*. Cittalia: Fondazione Anci Ricerche.

Atkinson, R. 2001. The emerging 'URBAN Agenda' and the European spatial development perspective: towards an EU urban policy? *European Planning Studies*, 9(3), 385–406.

Atkinson, R. 2007. *EU, Urban Policies and the Neighbourhood: An Overview of Concepts, Programmes and Strategies*. EURA Conference – The Vital City, Glasgow, United Kingdom, 12–14 September.

Atkinson, R. and Rossignolo, C. 2010. Cities and the 'soft side' of Europeanization: the role of urban networks, in *The Europeanization of Cities. Policies, Urban Change and Urban Networks*, edited by A. Hamedinger and A. Wolffhardt. Amsterdam: Techne Press, 197–208.

Bagnasco, A. 1977. *Tre Italie*. Bologna: il Mulino.

Balducci, A. 2000. Le nuove politiche della governance urbana. *Territorio*, 13, 7–15.

Barbanente, A. and Monno, V. 2004. *Changing Images and Practices in a Declining 'Growth Pole' in Southern Italy: The 'Steel Town' of Taranto* [available at: www.schrumpfende-stadt.de].

Barca, F. 2006. *Italia Frenata. Paradossi e Lezioni della Politica per lo Sviluppo*. Rome: Donzelli.

Bodo G. and Viesti G. 1997 *La Grande Svolta. Il Mezzogiorno nell'Italia degli anni Novanta*. Rome: Donzelli.

Borri, D. 1996. Puglia, in *Le Forme del Territorio Italiano*, edited by A. Clementi, G. Dematteis and P.C. Palermo. Rome-Bari: Laterza, 299–336.

Briata, P., Bricocoli, M. and Tedesco, C. 2009. *Città in Periferia*. Rome: Carocci.

Bricocoli, M. and Savoldi, P. 2010. *Milano Downtown. Azione Pubblica e Luoghi dell'Abitare*. Milan: Et al. Edizioni.

Cassano, F. 1997. *Il Pensiero Meridiano*. Rome-Bari: Laterza.

CEC. 1997. *Towards an Urban Agenda in the European Union*. Brussels: European Commission. COM(97)197, final, 06.05.97.

CEC. 1998. *Sustainable Urban Development in the European Union: A Framework for Action*. Brussels: European Commission. COM(98)605.

CEC. 2002. *The Urbact Programme 2002–2006*. Brussels: European Commission.

CEC. 2006. *La politica di coesione e le città*. Brussels: European Commission.

Cersosimo, D. and Donzelli, C. 2000. *Mezzo Giorno. Realtà, Rappresentazioni e Tendenze del Cambiamento Meridionale*. Rome: Donzelli.

Chorianopoulos, I. 2010. Uneven development and neo-corporatism in the Greek urban realm. *Analise Social*, XL(V), 739–56.

Cipriano, P. 2010. Nuova managerialità pubblica e 'questione amministrativa' meridionale. Il 'caso' Palermo. *Meridiana*, 68, 95–123.

Cremaschi , M. 2003. *Progetti di Sviluppo del Territorio*. Milan: Il Sole 24ore.

Cremaschi, M. 2005. *L'Europa delle Città. Accessibilità, Partnership, Policentrismo nelle Politiche Comunitarie per il Territorio*. Firenze: Alinea.

Cremaschi M. ed. 2010. *Politiche, Città, Innovazione. Programmi Regionali tra Retoriche e Cambiamento*. Rome: Donzelli.

Curti, F. and Gibelli, M. eds. 1996. *Pianificazione Strategica e Gestione dello Sviluppo Urbano*. Firenze: Alinea.

Demattteis G., Governa F. and Rossignolo C. 1999. *The Impact of European Programmes on the Governance of Italian Local Systems*. EURA conference European Cities in Transition, Paris, France, 22–23 October.

Donolo, C. 1999. *Questioni Meridionali*. Napoli: L'Ancora.

Fedeli, V. and Gastaldi, F. 2004. *Pratiche Strategiche di Pianificazione*. Milan: FrancoAngeli.

Felice, E. 2007. *Divari Regionali e Intervento Pubblico. Per una Rilettura dello Sviluppo in Italia*. Bologna: Il Mulino.

Governa, F. 2010. Competitiveness and cohesion: urban government and governance's strains of Italian cities. *Analise Social*, XL(V), 663–83.

Governa, F. and Saccomani, S. 2004. From urban renewal to local development. New conceptions and governance practices in the Italian peripheries. *Planning Theory and Practice*, 5(3), 327–48.

Hamedinger, A. and Wolffhardt, A. 2010. Understanding the interplay between Europe and the cities: frameworks and perspectives, in *The Europeanization of Cities. Policies, Urban Change and Urban Networks*, edited by A. Hamedinger and A. Wolffhardt. Amsterdam: Techne Press, 9–39.

Kunzmann, K. 2010. Medium-sized towns, strategic planning and creative governance, in *Making Strategies in Spatial Planning: Knowledge and Values*, edited by M. Cerreta, G. Concilio and V. Monno. New York: Springer, 27–45.

Laino, G. 2010. Costretti e diversi. Per un ripensamento della partecipazione nelle politiche urbane. *Territorio*, 54, 7–22.

Lascoumes, P. and Le Galès, P. 2004. Introduction, in *Gouverner par les Instruments*, edited by P. Lascoumes and P. Le Galès. Paris: Presses de la Fondation Nationale des Sciences Politiques, 11–44.

Le Galès, P. 1998. La nuova 'political economy' delle città e delle regioni. *Stato e Nercato*, 1998, 53–91.

Le Galès, P. 2006. *Le Città Europee. Società Urbane, Globalizzazione, Governo Locale*. Bologna: Il Mulino.

Le Galès, P. and Lequesne, C. 1998. *Regions in Europe*. London: Routledge.

Leontidou, L. 1993. Postmodernism and the city: Mediterranean versions. *Urban Studies*, 30(6), 949–65.

Marshall, A.J. 2004. Europeanisation at the urban level: local actors, institutions and the dynamics of multi-level interaction. *ESRC/UACES Study Group on the Europeanisation of British Politics and Policy-Making*, Sheffield, United Kingdom, 23 April.

Martinelli, F. 2005. *La Pianificazione Strategica in Italia e in Europa*. Milan: FrancoAngeli.

Masella L. 2009. Divagazioni sul Mezzogiorno. Ipotesi per una ricerca, in *Pensare la Politica*, edited by F. Giasi, R. Gua and S. Pons. Rome: Carocci, 479–91.

Mayntz, R. 1998. New challenges to governance theory. *Jean Monnet Chair PaperRSC*, 98(50).

Mazza, L. 2004. *Piano, Progetti, Strategie*. Milan: FrancoAngeli.

Meldolesi L. 1998. *Dalla parte del Sud*. Rome-Bari: Giuseppe Laterza.

Monno, V. 2010. Exploring the limits of the 'possible' and the value of the 'impossible', in *Making Strategies in Spatial Planning: Knowledge and Values*, edited by M. Cerreta, G. Concilio and V. Monno. New York: Springer, 161–83.

Pace, F. Forthcoming. Le politiche per la rigenerazione urbana e territoriale in Puglia: una iniziativa di successo. *Urbanistica Informazioni*, 243.

Padovani, L. 2010. Policy approaches and facts in Italy, in *Quality of Urban Public Spaces and Services as Paths Towards Upraising Deprived Neighbourhoods and Promoting Sustainable and Competitive Cities*, edited by L. Padovani. URBACT II, Sha. Keproject Baseline study, 37–46.

Palermo, P.C. 2001. *Prove di Innovazione*. Milan: FrancoAngeli.

Parkinson, M. 2005. Urban policy in Europe. Where have we been and where are we going?, in *European Metropolitan Governance. Cities in Europe-Europe in the Cities*, edited by E. Antalovsky, J. Dangschat and M. Parkinson. Vienna: Europaforum, 17–68.

Parlangeli, R. forthcoming. La programmazione economica urbana della città di Lecce: processo o progetto? La città che verra, la via della rigenerazione e della partecipazione pubblica. *Urbanistica Informazioni*, 243.

Pasqui, G. 2010. Un ciclo politico al tramonto: perché l'innovazione delle politiche urbane in Italia ha fallito, *XXIV Convegno SISP*, Venezia, Italy, 16–18 September.

Pichierri A. 1999. Dalle economie regionali alle economie urbane: la riscoperta della città. *Sviluppo Locale*, VI(12), 105–27.

Pichierri, A. 2002. Concertation and local development. *International Journal of Urban and Regional Research*, 26(4), 689–706.

Sabel, C.F. 1988. La riscoperta delle economie regionali. *Meridiana*, 3, 13–71.

Seixas, J. and Albet, A. 2010. Urban governance in the South of Europe: cultural identities and global dilemmas. *Analise Social*, XL(V), 771–87.

Tedesco, C. 2003. Europeizzazione e politiche urbane nel Mezzogiorno d'Italia, *Urbanistica*, 122, 49–54.

Tedesco, C. 2005. *Una politica 'europea' per la città? L'implementazione di URBAN a Bari, Bristol, Londra e Roma*. Milan: FrancoAngeli.

Tedesco, C. 2006. Territorial action and EU regional policy in the Italian Mezzogiorno: hybridizing 'European' frames in local contexts, in *Rethinking European Spatial Policy as a Hologram. Actions, Institutions, Discourses*, edited by L. Doria, V. Fedeli and C. Tedesco. Aldershot: Ashgate, 89–104.

Tedesco, C. 2007. Piani strategici in Puglia. *Urbanistica Informazioni*, 213, 13–14.

Tedesco, C. 2009. Innovation and 'resistance to change' in urban regeneration practices: a neighbourhood initiative in Southern Italy. *Journal of Urban Regeneration and Renewal*, 3(2), 128–40.

Tedesco, C. 2010. EU and urban regeneration 'good practices' exchange: from download to upload Europeanization?, in *The Europeanization of Cities. Impacts on Urban Governance and on the European System of Governance*, edited by A. Hamendiger and A. Wolffhardt. Amsterdam: Techne Press, 183–96.

Tedesco, C. 2011. Negli interstizi delle azioni 'innovative' di rigenerazione urbana. *Archivio di Studi Urbani e Regionali*, 100, 82–98.

Tofarides, M. 2003. *Urban Policy in the European Union. A Multi-Level Gatekeeper System*. Aldershot: Ashgate.

Trigilia, C. 1992. *Sviluppo senza Autonomia*. Bologna: Il Mulino.

Viesti, G. 1999. Sulle condizioni per lo sviluppo dei sistemi produttivi locali. *Meridiana*, 34–5, 33–70.

Viesti, G. 2003. *Abolire il Mezzogiorno*. Rome-Bari: Giuseppe Laterza.

Viesti, G. 2009. *Mezzogiorno a Tradimento. Il Nord, il Sud e la Politica che non c'è*. Rome-Bari: Giuseppe Laterza.

Chapter 5

The Challenges of Urban Renewal: 10 Lessons from the Catalan Experience[1]

Oriol Nel·lo[2]

In 2004, the parliament of Catalonia approved the *Law of Improvement of Neighbourhoods, Urban Areas and Small Towns Requiring Special Attention.*[3] This regulatory framework, which bears some singularities whether at the Iberian or the European levels, seeks to foster comprehensive renewal projects in those neighbourhoods experiencing the greatest urban deficits and where, as a result, populations needing social assistance tend to concentrate. In the subsequent seven years, comprehensive interventions took place in 141 neighbourhoods inhabited by over 1,000,000 people (13 per cent of the Catalan population) and investment amounting to 1,330 M€ was allocated, funded equally by both the regional government [*Generalitat de Catalunya*] and the respective municipalities, of which 513 M€ had actually been invested until December 2010 (Map 5.1).

The programme was implemented in the period 2004–10. From this date, the shift in the government of Catalonia and the effects of economic crisis have conducted a number of changes that, in fact, entail the cessation of new calls for projects and severe difficulties for the ones being implemented. In March 2012, the parliament of Catalonia voted to re-establish the programme in 2013. However, due to the circumstances several of the projects initiated during the period 2004–10 may not be completed.

Despite these problems, the study of Catalan case is of interest both from an academic and political standpoint. This is so because the application of this programme has brought to light some of the recurrent challenges to urban renewal processes: complexity in the physical intervention, financing problems, the need to comprehensively intervene, difficulties in achieving cooperation between

1 An early version of this chapter was published in 2010 in *Análise Social. Revista do Instituto de Ciências Sociais da Universidade de Lisboa*, 197, 685–715.

2 Universitat Autònoma de Barcelona; Departament de Geografia. 08193 Bellaterra (Spain). oriol.nello@uab.cat.

3 According to the Spanish Constitution and its Statute of Autonomy, Catalonia has exclusive competences in terms of urban and spatial planning. This ascription of competences also includes urban renewal policies. The full text of the *Ley de mejora de barrios, áreas urbanas y villas que requieren atención especial* can be found in *Diari Oficial de la Generalitat de Catalunya*, 4151, of 10 June 2004.

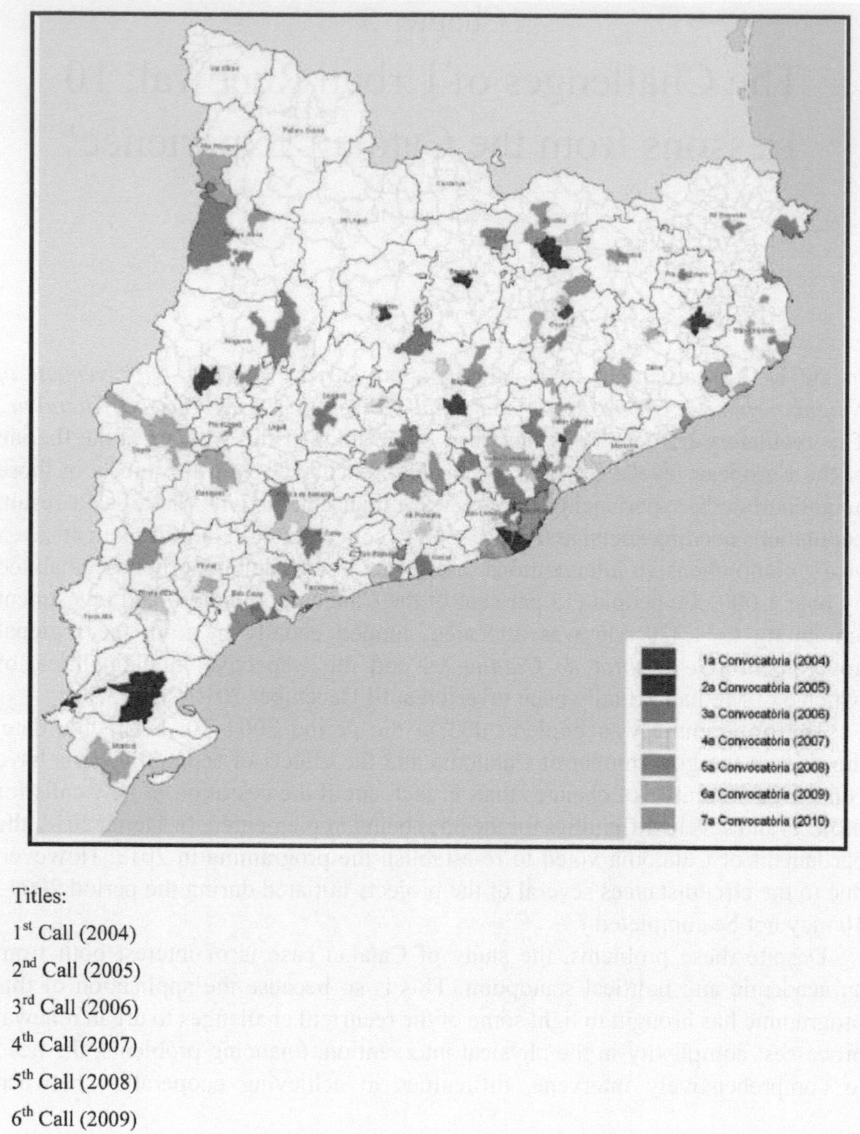

Titles:

1st Call (2004)

2nd Call (2005)

3rd Call (2006)

4th Call (2007)

5th Call (2008)

6th Call (2009)

7th Call (2010)

Map 5.1 Territorial distribution of the Law on Neighbourhoods programme (2004–10)

Source: Generalitat de Catalunya. Departament de Política Territorial i Obres Públiques

different administrative levels, active involvement of neighbourhood residents, and so forth. These challenges were met using tools we may classify as relatively innovative. Hence, the application of the Law on Neighbourhoods constitutes, in our opinion, an important case when studying the governance challenges posed within the context of urban transformation in Europe.

This chapter therefore provides an overview on this practical experience in order to reach broader conclusions.[4] As stated, these conclusions have been articulated in 10 proposals, lessons – disciplinary, administrative and political – to be drawn from this experience. The structure of the chapter responds to these 10 proposals, devoting a section to analyse each one.

A last warning before we begin. In this chapter, we adopt the word 'urban renewal' [*rehabilitación urbana*, in Spanish], a wording we prefer, instead of the concept 'regeneration', as the latter is pervaded with organicist and even moral meanings. In Spanish, the term *rehabilitación* has two meanings in the dictionary. The first – '*to return to its former state*' – would be inadequate for our purposes given that many of the neighbourhoods discussed have experienced serious inefficiencies right from the moment they were built. On the other hand, the second meaning – '*to re-establish its own rights*' – incorporates exactly the objective of that which, in our opinion, urban policies have to strive to achieve: to re-establish rights for all citizens irrespective of just where they live. Hence, it is in this sense that we subsequently employ *renewal*.

The Need for a Comprehensive Vision

As is known, the concentration of populations with lesser purchasing power and greater social needs in certain neighbourhoods is a direct result of the process of urban segregation, that is, the phenomenon through which, due to their respective options in the land and housing market, different social groups tend to become separated in urban areas. As the capacity to choose a place to live depends on both personal and household income, those with greater income obviously have greater liberty when it comes to choosing where they want to live, while households of lesser financial means are generally pushed to places where prices are lower.[5] Hence, social groups with lower incomes and higher social needs tend to group in these urban areas with greater urban deficits.

4 Data and statistical information used here were compiled for the studies carried out in 2009 evaluating this Law and its impact four years on from its enactment. These studies and all statistical evidence are found in Nel·lo (2009).

5 David Harvey (1973) wrote a classic text on the mechanisms involved in urban segregation processes and their results. In the case of the metropolitan area of Barcelona, the economist Carme Trilla (2002) and the geographer Francesc Muñoz (2004) quantified the capacity of different social groups to choose their residence comparing the average income of the population in each municipality and the respective average housing prices.

The causes and effects of urban segregation on the increase in social inequalities in the major Spanish cities have been thoroughly studied (García-Almirall et al. 2008, Leal and Domínguez 2008, López and Rey 2008, Nel·lo 2004). Firstly, they entail the paradox that municipalities with the biggest urban deficits and the greatest needs in terms of social services tend to have a more limited fiscal base, while areas where these needs are lower have more resources as a result of their capacity to levy fiscal charges. From a town planning perspective, this means that, generally speaking, neighbourhoods and towns where the population with greater purchasing power is concentrated tend to have better public spaces and better public equipment and attract better private services. Conversely, areas with lower income populations are where deficits are historically higher and home to higher demands on public space and public equipment and have greater difficulties in financing their acquisition and implementation.

We encounter similar effects within the housing market, especially in countries like Spain, where the rental market has a relatively reduced weight and where most families own the housing units they live in: neighbourhoods inhabited by lower income populations are those where, in principle, buildings tend to be older and poorer quality and hence the owners experience greater difficulties in maintaining them. Furthermore, the concentration of rather problematic social situations in such areas reduces expectations as to property values and this influences land owners when considering engaging in possible renewal projects.

Moreover, in many cases, shortages in terms of public spaces, infrastructures and housing are accompanied by high commuting costs, in terms of both economic resources and time, given the usually peripheral position of many of these neighbourhoods in relation to the centre of the respective urban area. Finally, we should also consider that segregated reproduction patterns amongst social groups, especially as far as education is concerned, may pose a significant barrier to equal opportunity and social mobility.

In Catalonia, however, the effects of urban segregation have been more benign than in other European countries and, since the restoration of democracy, there has been a considerable improvement in the standards of living of most districts and cities. The progress occurred due to the combined results of demand and efforts by local communities, the policies implemented by the public administration, overall economic development, and the dynamics of spatial integration, especially in the metropolitan area of Barcelona. Here, these dynamics have even resulted in a reduction in the cleavages in income distribution between the city and the rest of the area, falling consistently over the last two decades of the twentieth century (Giner 2002, Subirats 2012).

However, since the mid-1990s, the risks of rising social segregation have tended to worsen and, in certain places, some problems that seemed to have been overcome have re-emerged – such as housing overcrowding, degradation of public space, and difficulties as to the provision of basic services. This development has been brought about mainly by two factors: the evolution of the real estate market in Spain and the change in demographic trends (Nel·lo 2008).

Firstly, the *housing market* witnessed a rapid increase in prices, which began in 1996 and lasted for more than a decade and did not stop until 2007 (Figure 5.1). As a result of these market developments, the percentage of income families have to allocate to housing costs has increased dramatically, to the point that accessing affordable housing became difficult for much of the population.[6]

The rise, driven to a large extent by the financial sector, was further accompanied by a *significant inflection of demographic dynamics*. After a long period of stagnation, the Catalan population went from 6.2 to 7.5 million people in little more than a decade (1996–2009). This growth, as shown in Figure 5.2, was due largely to immigration inflows, which led to greater demand for housing, with the particularity that this demand came from a population segment which, in most cases, was barely solvent in comparison with prevailing market conditions. Other elements, such as the diminution of household average size and the arrival at the emancipation age of the large population cohorts born in the 1960s and 1970s, contributed as well to push the demand upward.

The combination of these two factors – the housing market and demographic changes – gave rise, on the one hand, to the re-emergence of substandard housing situations (especially as a result of overcrowding) and, on the other hand, to the concentration of social groups with lesser purchasing power in places where housing was relatively more affordable. As a result, the increased risk of social segregation and, in particular, the concentration of incidences of greater social need in those very neighbourhoods experiencing the larger urban deficits: the old centres, 1960s and 1970s housing estates and areas resulting from processes of marginal urbanisation.

The main objective of the Law on Neighbourhoods is to address these problems, avoiding the degradation of living conditions in these areas and acting, where possible, on factors at the root of urban segregation.[7] The aim was, firstly, to promote social and spatial justice, so that all citizens, regardless of their place of residence, may have fair and equitable access to basic services and a quality environment. With that, the Law tries to prevent urban dynamics from contributing towards greater inequality in opportunities. Simultaneously, the Law has the objective of enhancing the city through social justice, on the premise that

6 According to the survey compiled every year by the Catalan Government Environment and Housing Department Housing Secretary, in 2007 households with income equalling 3.5 times the minimum wage would have to allocate more than 123 per cent of their income to afford a new home in Barcelona; almost 90 per cent in the metropolitan area of Barcelona, and 66 per cent in the rest of Catalonia (Secretaria d'Habitatge 2009). See as well Donat (2010).

7 For further insight on the objectives and the potential of the Law on Neighbourhoods, see Muñoz (2006), Albors (2008), García-Ferrando (2008) and Nel·lo (2009, 2011). In order to place this Law within the context of renewing the spatial management tools in Catalonia, see Nadal (2007) and Nel·lo (2005, 2012). In order to compare the initiative with other experiences in Spain and other EU member countries, see Mongil (2010) and Van den Berg et al. (2007).

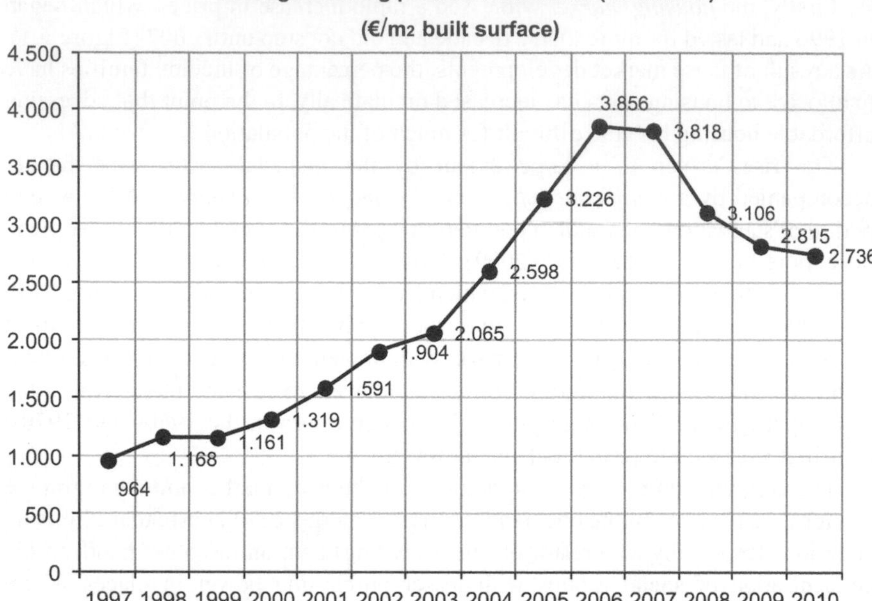

Figure 5.1 Average housing prices in Catalonia (1997–2010)

Source: Generalitat de Catalunya. Secretaria d'Habitatge. (Prices refer to new housing units.)

a city without social fractures is a more liveable, favourable and appealing space for all inhabitants (Indovina 1992).

Given these approaches, one could argue that problems relating to urban decline are primarily local issues and, hence, their treatment through regional policies is inadequate. However, greater spatial integration leads to a situation where the residential market in which citizens and private sector actors make their decisions is not their immediate surroundings, but covers a much wider scope. Nowadays, urban segregation does not arise only amongst neighbourhoods in the same locality, but also between districts within the same urban area and even the whole region. Segregation has become a phenomenon of metropolitan and regional scope. Hence, a comprehensive vision covering the entire region is needed in order to fund and implement urban renewal projects. Only based on such visions resources may be obtained and equitably distributed to districts and municipalities experiencing the greatest difficulties. Urban segregation in Europe in general, and Catalonia in particular, is no longer a local matter: it responds to social and spatial dynamics of metropolitan scale at least and should be opposed with the resources and will of the whole society. This is the first lesson drawn from the implementation of the Law on Neighbourhoods in Catalonia.

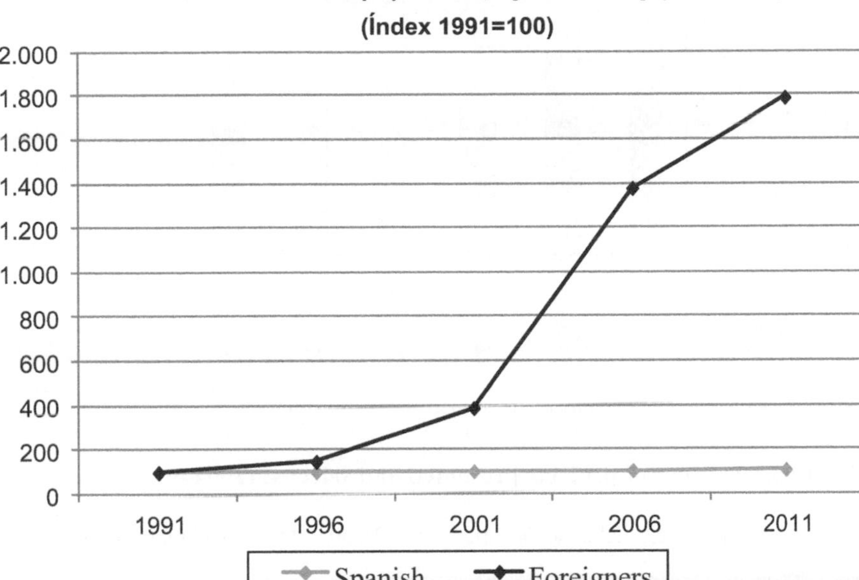

Figure 5.2 Catalonia resident population, by nationality (1991–2011)

Source: Statistic Institute of Catalonia.

Financing Projects, not Problems

The second lesson to be learned from implementation states that, when allocating resources, it is much more adequate and effective to target projects rather than problems. The legal mechanism established operates very simply and it is to an extent inspired by the European Union URBAN programme.[8] It is based on creating a financial fund by the Department of Territorial Planning and Public Works to which regional government resources are allocated. Based upon this resource pool, the government annually issues a call for municipalities seeking to carry out comprehensive rehabilitation projects in their neighbourhoods to submit applications. When approved, projects receive funding that varies from 50 per cent to 75 per cent of total project costs in accordance with the legislative framework.

This raises the key issue of how to select which projects are to be awarded financing, because as shown in Figure 5.3, the number of applications presented each year by far exceeds the regional government's funding capacities.

8 As to the URBAN mode and its methodology, which is termed the *urban acquis*, see the works by Gutiérrez (2008, 2009) and De Gregorio (2010).

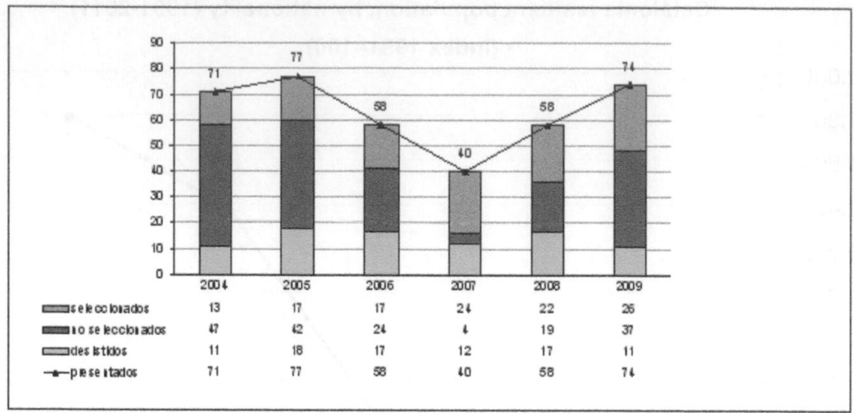

	2004	2005	2006	2007	2008	2009
seleccionados	13	17	17	24	22	26
no seleccionados	47	42	24	4	19	37
desistidos	11	18	17	12	17	11
presentados	71	77	58	40	58	74

Key: selected, non-selected, withdrawn, presented

Figure 5.3 Number of projects presented and selected (2004–9)

Source: Generalitat de Catalunya. Departament de Política Territorial i Obres Públiques.

From the outset, it might seem obvious that the selection criteria should be based essentially on statistical evidence of deficits and social deprivation experienced by neighbourhoods. However, international experience demonstrates that choices made solely on the basis of these parameters may end up being counterproductive. One factor contributing to worsening the objective difficulties of places affected by segregation is their stigmatisation by the media and also the subjective loss of self-esteem amongst their populations.[9] Under these circumstances, to establish and publish a ranking of the most needy districts might contribute to consolidate this negative image. Furthermore, the fact that a locality is in poor objective conditions does not guarantee, in itself, that the intervention project designed by the respective municipality is the most appropriate to tackle the existing problems.

For these reasons, implementation of the Catalan law made recourse to a dual system of tables when selecting projects. First, the position of the locality is evaluated according to 16 objective statistical indicators on four areas: urban and equipment deficits, population structure and dynamics, environmental problems, and economic and local development deficits.[10] In order to be considered a 'special

9 See, for example, in relation to the French case, Mucchielli (2007).

10 The 16 indicators, grouped by fields, are the following: a) Urban regression processes, under-provision of equipment and services (registered values, poor state of building conservation, buildings without water or sewage services, four-story housing buildings without elevators); b) Demographic indicators (population density, excessively rising or falling population levels, rates of dependence, levels of immigration); c) Economic, social or environmental problems (number of citizens receiving government aid and fully government-

attention area', the neighbourhood in question must return a minimum score from this set of indicators.

After this first consideration of the reality in the area under study, we move on to the second evaluation process phase, based on *the characteristics of projects* submitted by councils. In this second phase – which in terms of evaluation has equal weighting to that resulting from the socio-economic indicators – priority is given to projects based on criteria such as the consideration of its comprehensive character, its overall coherence, the economic engagement of the municipal authorities, citizen participation, the parallel undertaking of complementary actions, and others.

From merging the two groups of indicators – statistics and project related features – the score assigned to each proposal is obtained and, based on this score, available resources are then allocated by each call. On the one hand, the use of these objective parameters is there to avoid a possible perverse effect in the distribution of resources by a supra-local body: the risk that the allocation process is reduced to a competition between municipalities in need (Atkinson 2003). On the other hand, this analysis of the benefits and effectiveness of the project submitted by a town council is there to ensure allocated resources are used efficiently and in line with the legally stipulated objectives. Thus, it can be stated that the programme is not only a tool for neighbourhoods with problems, but a device for helping neighbourhoods with projects: projects conceived to address existing problems.

The Transversal Nature of the Intervention

A third lesson deriving from implementation of this law reflects the need for an integrated approach to urban renewal problems: only by intervening simultaneously on urban and social shortcomings can a policy be reasonably effective and consistent in the improvement of standards of living. This proposal achieves across-the-board agreement at least from the theoretical perspective (Cremaschi 2005, Gutiérrez 2008, Parkinson 1998), but has traditionally encountered great difficulties in practical application by administrative structures set up in accordance with the principle of the division of tasks and the sectoral budgetary distribution. In the implementation of the Law on Neighbourhoods, the intention was to encourage such a transversal approach by means of three instruments: the requirement for intervention across a number of fields in order to qualify for funding, the creation

funded pensions, high unemployment rates, lack of green areas, low levels of education, social and urban deficits); d) Local development problems (lack of public transportation, lack of parking places, low levels of business activities, percentage of citizens at risk of social exclusion). See decree *369/2004, de 7 de septiembre, por el que se desarrolla la Ley 2/2004, de 4 de junio, de mejora de barrios, áreas urbanas y villas que requieren atención especial* (*Diari Oficial de la Generalitat de Catalunya*, 4215, de 9 de septiembre 2004).

of complementary programmes by various Generalitat departments, and the establishment of integrated monitoring mechanisms.

Regarding *fields of intervention*, the law sets out eight different areas in which action can be taken by municipalities covering in this way a wide diversity of fields. It is worth recalling their reach and ascertaining the scopes prioritised by the city councils when allocating investment (Table 5.1).

- *Improvement of public spaces and provision of green areas* (road paving, placement of trees, lighting, landscaping).
- *Renovation of residential buildings* (façades, gutters, elevators, roofs).
- *Provision of collective facilities* (civic centres, elderly homes and senior centres).
- *Installation of communication technologies* (wiring of buildings, establishment of Wi-Fi areas).
- *Implementation and improvement of energy and environmental infrastructures* (pneumatic waste collection, underground containers, opening recycling centres, encouraging renewable energies, water-saving devices).
- *Fostering gender equity in the use of urban spaces and facilities* (introduction of a gender perspective in the provision, design and configuration of public spaces and facilities).
- *Development of programmes targeting social, economic and urban improvement of districts* (support activities for groups at risk of social exclusion, training programmes, commercial revitalisation.)
- *Improving accessibility and the removal of architectonic barriers* (extension of sidewalks, construction of ramps, installation of escalators, and removal of obstacles).

Table 5.1 Investment by field of action 2004–9

Field	Total cost of the projects	Total cost of the projects (%)	Total funding
1. Improvement of public spaces and provision of green	514,762,162.1	44.41	262,135,715.3
2. Renovation of common areas in residential building	115,411,723.5	9.96	58,766,084.7
3. Provision of public facilities	252,999,297.0	21.83	130,211,400.5
4. Installation of information technologies into buildings	14,805,205.9	1.28	7,691,113.9

Field	Total cost of the projects	Total cost of the projects (%)	Total funding
5. Implementation and improvement of urban environmental sustainability: energy and water	54,944,949.6	4.74	28,781,634.9
6. Fostering gender equality in the use of urban spaces and facilities	24,666,118.1	2.13	12,696,463.6
7. Development of programmes targeting social, economic and urban improvements	108,853,190.0	9.39	56,286,073.6
8. Improving accessibility and the removal of architectonic barriers	72,496,677.2	6.26	37,281,513.1
Total	1,158,939,323.4	100.00	593,839,998.6

Source: Generalitat de Catalunya. Departament de Política Territorial i Obres Públiques.

In order to enhance the programme's transversal nature, the law and its regulatory framework stipulate that the score obtained for each project submitted in the selection process be proportional to the number of fields of action. This has led to the result that, out of the 92 projects selected in the first five calls, 71 intervene in all eight possible fields, eight in seven and seven in six. The least frequently included field of intervention (the promotion of new technologies) is present in a total of 79 projects. Indeed, no matter how transversal, a tool local in scope and endowed with limited resources may not prove to be a decisive influence on the evolution of structural variables – such as the labour market, migratory flows and the housing market – which should be handled with the appropriate economic and fiscal policies. However, the level of transversality achieved by the Law on Neighbourhoods may represent a significant advance towards avoiding a situation where territorial and urban issues exacerbate social problems arising from the general evolution of these variables.

In efforts to cover all material aspects of neighbourhood life, complementary actions provided by other regional government departments[11] proved very important. Worth mentioning among them is the 'Jobs in Neighbourhoods' programme run by the Department of Labour through the Employment Service of Catalonia, which resulted in the establishment of agreements in 81 of 92 districts covered by the law and providing them with occupational training and school-

11 For data pertaining to the evolution of these complementary programmes in the period between 2004 and 2008, see Amill et al. (2009).

to-work transition programmes with the commitment of 30 M€ total investment between 2004 and 2008. The Department of Health has conducted studies on the status of public health and healthcare in the districts through the 'Health in the Neighbourhood' programme, having intervened in 30 districts in the first three programme calls. Furthermore, the Department of Environment and Housing has set up specific lines of support for the renewal of collectively shared building areas for 37 districts, and the Departments of Home Office, Institutional Relations and Public Participation have financed *citizen participation processes* in a further 24. Finally, the Catalan Land Institute has signed agreements to carry out *urban renewal* activities (in particular, the replacement of obsolete housing with new units) in 24 programme districts with investment, in parallel to that generated by the law, estimated at 200 M€. Less successful was the coordination of actions undertaken in the fields of Education, Social Action and Immigration.

The institutional expression of this transversal approach is the creation of *an Evaluation and Monitoring Committee* in each neighbourhood benefiting from the programme. These committees are the coordinating body for each project and are made up of, on the one hand, representatives from seven Generalitat departments (Territorial Planning and Public Works, Environment and Housing, Governance, Social Action, Economy and Finance, Health and Labour, and the delegation of the regional government in the respective province), and, on the other, the city services directly responsible for the district management. Thus, the Evaluation and Monitoring committees become bodies with great potential when it comes to bringing together, in the presence of the respective mayor, all Generalitat and municipal council services related to local life. However, precisely because of their complexity, the committee meetings cannot be very frequent (on average once per year in each district) and do not replace daily Generalitat and municipality activities. The Law on Neighbourhoods thus sets out a core infrastructure for integrated and transversal action in the urban areas within this special programme but the process of administrative adaptation in order to act more territorially and less sector-oriented (in neighbourhoods requiring special attention and in many other areas) still requires a great deal of further change (Amill et al. 2009).

The Key Role of Public Investment

The fourth lesson derived from the implementation of the Law on Neighbourhoods is the acknowledgement of the key role that public investment plays in urban renewal processes. As discussed above, under current market conditions, the most difficult social situations tend to cluster where prices are lower. This leads, firstly, to difficulties in the provision and access to basic services, and, secondly, to increasing risk that average house prices in this area tend to move further and further from those of the city as a whole. In this context, investment and government spending on districts, besides having a direct impact on social problems and urban deficits, becomes a catalyst for attracting more investment from other sources (public or private).

The first immediate effect of the Law of Neighbourhoods[12] was to provide *greater resources for tackling social problems* specific to each neighbourhood. This enabled authorities to provide special attention to those groups that are most exposed to risks of exclusion (the elderly, young and recent immigrants), as well as to promote jobs, commercial activity and foster gender equity in access to services and public spaces. The resources rendered for these purposes by the programme, 134 M€ of expenditure incurred in the first five years of implementation (Table 5.1, rows 6 and 7), come in addition to (and do not replace) regular budgets allocated by the government and municipalities for social provisions in general in these very same districts (unemployment benefits, social subsidies and others). Thus, programme areas gain additional resources, not only to provide direct aid to individuals, but also to foster community programmes that promote equal opportunities in access to income and services across the whole of the district.

A second effect of public investment in the neighbourhoods covered by the programme has been to *reduce deficits in town planning, equipment and housing conditions* currently witnessed in these areas. These deficits represent a significant obstacle to quality of life and decisively impact the housing price formation mechanism and the resulting dynamics of segregation. In the first years of programme implementation, municipalities allocated a substantial share of available resources to this purpose. Thus, as seen (Table 5.1, row 1), almost half of the investment committed (515 M€) has been designed to meet needs relating to upgrading public spaces and providing green areas. This comes as no surprise: in neighbourhoods requiring special attention, public spaces not only experience deficits from the time of their very opening, but often are under great pressure (because their available area is often limited and because residents have little space indoors). Their expansion and recovery is therefore essential to improving the quality of community life and to tackling the commonly encountered conflicts over the use of collective spaces.

The investment of 55 M€ in the *improvement of environmental infrastructures* (waste collection and treatment, the water and energy cycles) pursued the same goal of reducing deficits that impair standards of living and equalising the status of all the neighbourhoods of the same municipality. This also applied to the investment in *facilities for collective use* (253 M€). However, a non-negligible additional aspect is worth noting: many districts selected by the programme not only lack facilities for their own inhabitants, but also amenities and services attractive to the city as a whole. This feature, found especially in the mass housing estates and areas arising from marginal urbanisation, has deprived these neighbourhoods of centrality and converted them into areas seldom visited by non-residents. The Neighbourhood Programme seeks to remedy this situation through the location of facilities in

12 For a detailed analysis by field of investment under the auspices of this law, see Mier and Botey (2009).

these neighbourhoods which, in addition to serving their neighbours, are used by the entire city, such as the Frederica Montseny Civic Centre, in Manlleu, or the renovated Coma-Cros factory, in Salt. This has helped to attract other public and private facilities, whether municipal, county or even of regional reach (such as the district courts in Balaguer or the Palace of Justice and the Fundació Catalana de l'Esplai headquarters in the district of Sant Cosme in Prat de Llobregat).

Together with the direct effects of the neighbourhood investment programme described above, there are other indirect effects. First, the existence of the Law on Neighbourhoods framework generated a catalyst effect in attracting *complementary public investment* to these neighbourhoods, either by the council itself, other Generalitat departments (some of which have created, as seen, their own specific binding programmes), and the central state. It is still too early to quantify the scope of this catalysing effect but early estimates (Picorelli and Saborit 2009) based on the experience of neighbourhoods in the first call, indicate that each euro allocated under the auspices of the Law on Neighbourhoods achieved a multiplier effect of 4.41. Thus, from the 179.7 M€ allocated by the law to the 11 first call districts (excluding the City of Barcelona), 793.6 M€ in public investment was committed.[13]

Furthermore, the concentration of this volume of public investment, either directly from the programme or attracted precisely due to it, has had a powerful effect on *fostering private investment*. Indeed, solving the built environment and facilities deficits is a prime incentive for private investment in neighbourhoods, especially by property owners. These, in a context of neighbourhood shortcomings and deteriorating standards of living, would have little incentives to renovate their own properties and would tend more to sell, even at low prices. With the implementation of the law, owners see the situation reversed as a result of public investment and act accordingly. Aggregate data are difficult to obtain on the total amounts of such investment, but there are very significant indicators relating to private sector investment renewal activities (Muñoz et al. 2009, for data see Table 5.2).

Let us, for example, consider the municipalities to which the neighbourhoods benefiting from the first call belong (excluding the City of Barcelona). In them as a whole the number of rehabilitated housing rose from an index of 100 in 2004 to 394 in 2007, whereas in the very areas of neighbourhoods benefiting from the programme growth soared from 100 to 668. It is also significant that, in spite of the trend of new building to concentrate in new, developing areas of the city and not in already consolidated districts like the programme ones, new housing dynamics do not greatly differ in their respective municipalities' average. Thus, while in indexed numbers the new dwellings completed in all municipalities (again excluding Barcelona) rose from 100 to 438, neighbourhoods subject to

13 The estimates should be viewed with caution, since they are derived from data provided during the first half of 2008 by the municipalities responsible for implementing the programme. These are, therefore, rather indicative data which, on the one hand, may have underestimated the initial provisions and, on the other, overrated the actual investment.

renewal processes jumped from 100 to 398. Additionally, if we consider not only effectively rehabilitated or completed homes, but also construction expectations (expressed as construction or rehabilitation licenses and permits), in none of the districts analysed did the expectation of housing production slide or register any decline between 2004 and 2007, contrary to what took place in some of their respective municipalities.

Given these dynamics, it may be argued that, depending on market conditions, such processes may lead to rising property prices and to fostering gentrification processes.[14] One could think that this situation would affect urban centres especially, more than housing estates or spaces emerging out of marginal urbanisation, as these are, fortunately or unfortunately, areas far from being exposed to this problem. However, this risk has been prevented in the Catalan case due to the introduction of policies to ensure social housing within any substantial new urban development.[15] Another factor has contributed to avoiding this risk: in the neighbourhoods benefiting from the first programme call (excluding areas within the city of Barcelona) 239.95 M€ of public investment went into housing. Thus, the investment committed to social housing in these areas exceeded even the resources allocated under the Law on Neighbourhoods (Picorelli and Saborit 2009).

Furthermore, it should be noted that, in Catalonia, as in the whole of Spain, property ownership is the dominant form of tenure, with more than 80 per cent of households owning their own housing unit. This does not always have positive effects in terms of housing access but has a welcomed side effect in terms of renewal: whenever there are improvements in the neighbourhood, residents themselves see that it adds value to their property and therefore share an objective interest in improving it. The predominance of ownership over renting also leads to deeper roots being set down in the neighbourhood, thus inhibiting the tendency of families who experience an increase in their level of income to relocate to other areas of the city. This clearly differentiates the Catalan experience from those cases where urban renewal policies were implemented in areas where rental is the predominant property type, whether private (in which case we do in fact witness transfer processes in the resident population), or public (in which case, tenants obviously have no direct financial interest in the renewal of the homes where they reside).

---

14 For a critical vision on the relationship between urban renewal policies and gentrification see Smith (2006).

15 This measure was established by the law – *Ley 10/2004, de 24 de diciembre, de modificación de la Ley 2/2002, del 14 de marzo, de urbanismo, para el fomento de la vivienda asequible, la sostenibilidad territorial y la autonomía local.* The law made it compulsory to all municipalities to devote a minimum of 20 per cent of each new housing development to protected housing. This obligation increases to 30 per cent in municipalities with more than 10,000 inhabitants and county capitals. These allocations are to be distributed between different modalities of protected housing, and balanced out across the whole of the city. For further details on the relationship between the Law on Neighbourhoods and Catalan housing policies, see Nel·lo (2008).

Table 5.2 Building, construction and renewal activities in Law on Neighbourhoods first call municipalities and neighbourhoods between 2004 and 2007

	All municipalities (excluding Barcelona)				All neighbourhoods (excluding Barcelona)			
	2004	**2005**	**2006**	**2007**	**2004**	**2005**	**2006**	**2007**
Finished residences (absolute values)	6,126	12,595	19,276	26,823	611	1,008	1,629	2,432
Rehabilited residences (absolute values)	562	1,114	1,800	2,214	56	170	276	374
Finished residences (accumulative values)	100	206	315	438	100	165	267	398
Rehabilited residences (accumulative values)	100	198	320	394	100	304	493	668

Source: Muñoz et al. (2009).

A New Type of Relationship between Regional and Local Government

The fifth lesson confirmed by the Law on Neighbourhoods: it is no longer possible to develop a wide reaching urban policy without intense exercise of inter-administrative cooperation. Detailed above are the reasons explaining why the intervention of supra-local levels of government in urban regeneration processes is essential. If the neighbourhood renewal has social cohesion as its main objective, then this necessarily must comprise some sort of public revenue transfers between different areas and hence the intervention of both higher and local authorities is necessary.

Thus, the Generalitat's presence is required for fostering, selecting, funding and evaluating projects, but it would be inappropriate for the regional government to try to perform such actions on its own: proximity is a prerequisite for urban policy success. Correspondingly, trying to intervene in the local arena directly from higher bodies central or regional can often lead, as the European experience has shown, to errors in assessment and action (Gutiérrez 2008, Parkinson 1998).

That is why this law has sought to assign full responsibility for project implementation to the municipalities. Thus, the city council – the institution with direct knowledge of the problems and potentials of each neighbourhood – is charged with designing and monitoring project implementation. From this perspective, the neighbourhoods programme seeks stricter compliance with the principle of subsidiarity and may be considered, since its enactment, as an essentially municipalist project. A project where the regional government relies on councils to promote and finance projects, and explicitly eschews any claims to play the main part.

This spirit of cooperation between levels of government has extended not only to programme design and implementation but also the very process of project selection. Thus, the law attributes responsibility for the selection and allocation of resources to the *Comisión de Gestión del Fondo de Fomento de Barrios y Áreas Urbanas que requieren una atención especial* [Management Fund Committee for the Development of Neighbourhoods and Urban Areas]. This committee, consisting of a total of 30 members, is equally made up of representatives from various Generalitat departments and local government associations (the Catalonia Federation of Municipalities and the Catalan Association of Municipalities), as well as professional architect and technical specialist orders. It is worthy to note that in seven calls for projects, the Commission distributed resources amounting to 1,330 M€ to 141 projects, with decisions always having to be taken unanimously. This might be proof that the system of tables and parameters put into effect by the law was well designed and appropriately subject to implementation. Furthermore, it also clearly signals the desire for transparency and multi-level governance that has been one of the main features of this programme.

As stated above, enforcement of this law is meant to complement inter-administrative cooperation of a vertical nature with horizontal intra-administrative cooperation. The latter is an essential prerequisite to achieving a transversal approach to policies and has often resulted in great complexity, within the Generalitat as well as in the municipalities. As noted before, the main expression of both inter-administrative and intra-administrative cooperation is the *Comité de Evaluación y Seguimiento* [Appraisal and Monitoring Committees] in each neighbourhood, which gathers into a single body all the regional and local administration departments with responsibilities for local life.

Potentials and Challenges of Sharing Finance

The shared funding model that characterises the programme corresponds to this willingness to use public investment as a lever to transform neighbourhoods. The importance of and difficulty in jointly managing the funding is the sixth lesson derived from implementation of the law.

As mentioned above, the funding mechanism is at the heart of this legislative framework and works simply: the Generalitat creates a financial fund for neighbourhoods from which municipalities receive project funding, which in the first five calls amounted to 50 per cent of the total planned investment and in the sixth and seventh calls was raised in some cases to 75 per cent. However, the apparent simplicity of this approach involves considerable complexity when it comes to application. The difficulties have three origins: the adequacy of resources, administrative management, and the atypical nature of planned spending.

Regarding the *adequacy of resources*, it should be noted that programmes under the auspices of this legislation entail an economic commitment which is highly significant for all authorities involved. Between 2004 and 2010, the Department of Planning and Public Works allocated 99 M€ to each call to be borne by the relevant multi-annual expenditure. This is undoubtedly a substantial financial sum but, in relative terms, the commitment is even higher for the vast majority of municipalities involved in the programme. Indeed, programmes in the first five calls had an average budget of 10.7 M€. This means that municipalities must make an average contribution of more than 1.3 M€ per year and neighbourhood. Thus, municipalities involved in the programme had to allocate, for each of the four operational years, contributions ranging from 22.5 per cent (in municipalities with fewer than 10,000 inhabitants) and 2.24 per cent (in those municipalities with between 100,000 and 500,000 inhabitants) of their annual budgets.

The efforts of council authorities participating in the programme are thus self-evident. Moreover, in the first year of their respective project municipalities have been able to make a very small percentage of the investment initially planned. No wonder, then, that there have been delays – in some cases quite substantial ones – in the execution of projects. Thus, the execution time of four years initially planned for the implementation of each project had to expand in almost all cases by one or two years, a possibility already provided by the law.

With the beginning of the economic crisis in 2008 some voices expressed doubts as to the ability of Catalan municipalities to continue meeting their commitments. However, as explained below, these pessimistic forecasts did not prove accurate at first. Despite the difficulties, council authorities honoured, in general, the commitments made both to citizens and to the Generalitat as to the implementation of neighbourhood programmes until the end of 2010. However, from 2011, the new Catalan government decided not to make calls for new projects, claiming lack of funds available due to the economic crisis. Furthermore, in 2012, budget provisions have not been made in order to cover the costs associated with the implementation of the Law of Neighbourhoods. In March this year the parliament of Catalonia adopted a resolution calling for the reactivation of the programme by 2013. However, it is very likely that under these circumstances some projects may suffer considerable delays and some municipalities may opt to discontinue them.

Along with problems in resource adequacy, we must take into account the *complexity of the shared financing mechanisms*. In summary, the funding circuit involves the following steps: first, once the project is selected by the Fund Management Committee, the council receives a notification as to the provision of aid from the Generalitat. Second, based on the list and the scope of planned actions in each neighbourhood, the Generalitat and the municipality establish an agreement that includes an economic and financial plan detailing projections for the implementation of measures over the four-year programme life cycle and the annual contributions that each institution is to make to render them viable. Third, the municipality starts the tender process and the implementation of measures and, as it proceeds, certifies the amounts deployed to the regional government. Fourth, the regional government checks the performance of these actions and the accuracy of forecast expenditure before then transferring payment to the municipality.

It should be noted that the attempt to introduce European funding into this circuit failed to a certain extent. Indeed, the Generalitat sought to allocate part of the European Regional Development Funding (ERDF) that it manages to meet the financial needs of the programme. Although the amount initially considered for this purpose amounted to only a small part of the total resources (33 M€ of eligible expenditure, of which 16.5 M€ was to have been ERDF funded), its inclusion forced the adoption of a number of complex and stringent new regulations that have affected the financial management of the entire programme. For smaller municipalities, these regulations have proved rather complex and sometimes even retarded progress. This forced the rescheduling of eligible expenses so that at the end of the 2004–8 period, the actually implemented ERDF Programme European funds totalled only 5.5 M€.

Finally, according to its design, funding for comprehensive renewal projects involves a form of budgetary planning and execution that is atypical in many of the municipalities involved. Municipal budgets are traditionally structured in keeping with sectoral lines, covering annual periods, and applying generically to the whole municipality. However, the implementation of the law entails the need to integrate resources from various departments and municipal services, to adapt to non-annual schedules, and to concentrate in a particular area. Not surprisingly, several municipalities have encountered difficulties in adapting their structures and switching from sectoral, annual and generic budget criteria to transversal, programmatic and geographically circumscribed approaches.

However, despite the complexity of funding mechanisms, the use of resources between 2004 and 2010 worked in a fairly satisfactory manner. Thus, as of 31 December 2010, investment and total expenditure justified by the municipalities to the Department of Planning and Public Works was 513.2 M€. This reflects how all municipalities benefitting from the first call have justified as a whole 94 per cent of the planned spending. In fact, 11 out of the 13 districts of this call had completed their entire projects by this date. As regards the projects of the second call, 68 per cent of the planned investment had been implemented and five of the 17 districts included in it had finished their projects. The level of investment in December 2010 of the projects included in the other calls can be seen in Table 5.3. Therefore, although a good proportion of programme projects was carried out in five or six years, instead of four years, the level of Law on Neighbourhoods financial implementation in the period 2004–10 can be advantageously compared to rates of investment compliance attained across public administration of Catalonia and Spain.

Table 5.3 Invested resources to December 2010 (neighbourhoods included in the 2004–8 calls)

Call	Total resources for each call	Total resources allocated as to December 2010	Percentage of invested resources to December 2010 with respect to planned spending
2004	197,698,236.52	186,303,191.54	94
2005	198,000,000.00	135,255,646.26	68
2006	198,000,000.00	95,983,933.73	48
2007	198,000,000.00	57,405,415.06	29
2008	198,000,000.00	31,332,884.44	16
2009	169,239,323.68	6,169,488.67	n/a
2010	171,627,696.03	771,418.29	n/a
Total	1,330,565,256.03	513,221,977.99	n/a

Source: Generalitat de Catalunya. Departament de Política Territorial i Obres Públiques.

Local Community Involvement

In the same way that the intervention of an administrative body alien to the local reality presents, as previously stated, many problems when dealing with neighbourhood problems, experience has shown that renewal projects are hardly likely to achieve their objectives without the involvement of local communities. This is the seventh lesson drawn from the implementation of urban renewal in Catalonia.

As is known, the existence of urban deficits and social problems often entails a lack of social interaction, difficulties in conviviality, and self-esteem problems for residents. These circumstances make it necessary for the development of urban renewal projects to prioritise the empowerment of local citizens when deciding on how to advance the area where they reside (Atkinson 2003, Parkinson 1998). To achieve this, neighbours need to be involved in the definition, implementation and evaluation of urban renewal programmes. Only in this way can intervention become a collective project *by* the neighbourhood (and not just *for* the neighbourhood), a project that, if necessary, neighbours may require the authorities to accomplish.[16] In this way, far from being mere beneficiaries, local communities become activists in the neighbourhood improvement process.

Based on this premise, the Law on Neighbourhoods provides for the participation of representatives from neighbourhood bodies and associations in the monitoring and evaluation committees formed in each district. This allows them to be involved directly in monitoring and evaluating results at the highest level, i.e. in the body that controls due programme compliance. However, experience has also shown that resident programme participation should not be reduced to this scope.

This is so, first, because each neighbourhood committee is formed once the resources have been allocated and execution of the programme has begun. Correspondingly, beginning at this moment, local community participation would be excluded from the crucial process of project design. Given this eventuality, it should be noted that those municipalities that included resident participation right from the outset of programme definition have achieved excellent results. In this regard, the possibility was raised that future programmes may have a 'year 0', i.e. a period in which, once resources are allocated to municipalities, before initiating proceedings, they prepare, define and design programme content in conjunction with neighbours.

Second, the need to expand citizen participation beyond evaluation and monitoring committee meetings derives from the very nature of this body. In fact, as mentioned, given its wide and heterogeneous nature, these committees meet only once a year. Thus, neighbourhoods with greater community involvement are those where other permanent participatory mechanisms between neighbourhood organisations and the municipality have been created. Likewise, very encouraging results are obtained in neighbourhoods where, together with the ongoing participation process, consultation took place with the specific objective of defining, for example, the design of a square or a street.[17]

16 As to the appearance of neighbourhood movements in the major Spanish cities during the final decades of Franco's dictatorship, see the works by Borja (1986) and Castells (1983). As to its evolution in the last 40 years, see Domingo and Bonet (1998) and Pérez and Sánchez (2008).

17 For an assessment of participatory processes linked to the Law of Neighbourhoods, see Martí-Costa and Parés (2009).

Finally, we should point out what might seem obvious. The process of resident involvement requires, of course, the willingness of governments to act with transparency and to engage in dialogue through to the end of the process. But the richer the neighbourhood social life, the greater the organisation of neighbours, the more successful this process will be. Therefore, the presence of a strong associational life and a representative local community movement is another requirement for the neighbourhood involvement process in order to ensure that the local improvement proves successful.

The Importance of Capitalising on Experiences

The implementation of the Law on Neighbourhoods has led to innovations in the design and implementation of policies and in various areas of Catalan public administration. This has meant that successes and failures in programme implementation represent a collective learning process. The need to capitalise on this experience to improve policies is the programme's eighth lesson.

Indeed, since 2004, a remarkable body of knowledge has accumulated in teams that, from the Generalitat to the municipalities, have been involved in the management of such programmes. Likewise, private consultants and institutions have worked together with municipalities in the development and implementation of projects and have gathered a body of highly valuable knowledge as well.[18]

Furthermore, the technical staff in charge of programme coordination – about 100 across all of Catalonia – represent a new professional profile non-existent in Catalan municipalities before, with the exception of some large municipalities. Taken together, this body of knowledge constitutes a valuable social capital that has enriched the experience accumulated in urban renewal policies in Catalonia since the return of democracy in the late 1970s.

In order to capitalise on this knowledge, the Catalan government decided at the beginning of 2007 to promote an integrated network for all municipalities participating in the programme with the aim of sharing and evaluating experiences. This scheme of cooperation, called 'Network of Neighbourhoods with Projects', seeks to articulate and strengthen relationships of a more or less informal nature that have been woven between the many institutions and technicians involved in programme implementation. The aim was to avoid fragmentation and isolation amongst local actors, something witnessed in urban renewal programmes in other European countries (Atkinson 2003).

The network includes elected officials and municipal technical staff, as well as technical specialists from the Generalitat (from both the Department of Planning and Public Works, and other departments working together on the programme), technicians from other local authorities (in particular the provincial councils),

18 See Ysa (2011) for an analysis of the management experience of the neighbourhood programme.

private consultants, neighbourhood associations, and researchers engaged in academic work on issues relating to the Law on Neighbourhoods implementation processes.

The network operates four thematic working groups (Public Spaces, Accessibility and Sustainability; Facilities and Residential Building Upgrading; Social and Gender Programmes, Economic and Commercial Promotion; Citizen Involvement) whose coordination and dynamics are the responsibility of members of staff hired by the Department of Planning and Public Works. Each working group organises a series of seminars and workshops open to all network members, while also making visits to specific operations.

The results of network activities include the creation of a catalogue of good practices so that all participating municipalities and those wishing to join can source a list of successful experiences, along with the regular publication of a digital newsletter with information and a website with the relevant calendars, blog, forum, links directory and file repository.[19] One intangible, but not insignificant, result is the fact that the existence of the network has played an important role in nurturing a certain sense of community amongst people and institutions responsible for promoting and implementing programmes.

However, the willingness to cooperate has not stopped within the boundaries of the region, but has also extended to other regions of the European Union. In fact, as already mentioned, since its inception the Law on Neighbourhoods was largely inspired by the experience of the European URBAN Programme and, despite the fact that resources from this programme have long since disappeared, and other European funds proved unsuitable for programme purposes, the Catalan government has seen fit to promote different cooperation experiences within the European framework. These initiatives have a threefold purpose: to promote the interest of European institutions in urban renewal policies, capitalise on experiences in other regions and European states in this field, and disseminate the lessons of the Law on Neighbourhoods (Mier and Botey 2009). Thus, within the URBACT EU initiative framework, the Generalitat has promoted and led the creation of two successive working groups on urban renewal (called CIVITAS and NODUS), consisting of European regions and cities involved in this field in Italy, Poland, the United Kingdom, Romania, the Netherlands, Belgium and Spain.[20]

Another key aspect for conceptualising and capitalising on the experience of the Law on Neighbourhoods has been the attention received from universities and

19 The Law on Neighbourhoods website, with the bulletin, the catalogue of good practices, and the programmed activities can be found at http://ecatalunya.gencat.net/portal/faces/public/barris.

20 CIVITAS Working Group: *CIVITAS. A Regional Approach: Added Value for Urban Regeneration* and NODUS Working Group: *Linking Urban Renewal and Regional Spatial Planning*. URBACT Programme: the documents can be found at: http://urbact.eu/projects/civitas/documents.html and http://urbact.eu/fileadmin/Projects/NODUS/outputs_media/NODUS_Final_Report_def_01.pdf.

Table 5.4 Some indicators on compliance with Law on Neighbourhood objectives in first- and second-call municipalities

	Total scoring of indicators ascertaining whether an urban area requires special attention (%)		Do you believe life in your neighbourhood has improved over the last five years? (%)			If you could, would you move elsewhere? (%)	
	Original situation	**Current situation**	**Yes**	**No**	**Does not know**	**Yes**	**No**
First-call neighbourhood averages	48.37	44.79	51.3	32.8	15.9	14.3	85.7
Second-call neighbourhood averages	36.61	36.55	52.1	34.1	13.8	26.3	73.7

Source: Cardona et al. (2009).

professional orders, in such a way that the neighbourhood programme has been the subject of attention and study by several universities in Catalonia and abroad. Proof of the interest in the initiative across both professional and academic fields is the award received at the International Union of Architects at its 23rd World Congress held in Turin in 2008 and granted to 'the Government of Catalonia for its pragmatic approach to urban renewal and urban development within the existing urban fabric'.

The Commitment to Evaluating Results

Evaluation of the implementation and outcomes of public policies is a challenge that, traditionally, Catalan and Spanish administrations have given only a limited response to. The need to address this issue in order to improve future programme development and as a precondition of rigour and transparency is the ninth lesson from implementing the Law of Neighbourhoods.

In fact, the same law did indeed incorporate its own assessment mechanisms. These are of two types: the first is assigned to the Evaluation and Monitoring Committee in each district, which is held responsible for drafting a 'final evaluation report'. Second, the Generalitat of Catalonia is required, after an initial four-year period of enforcement, 'to make an assessment of the objectives achieved and the needs of continuity, without prejudice to the ongoing proceedings'.

Regarding the first commitment, the Generalitat established that each *final evaluation report* was to contain a performance assessment of results in terms of urban quality, economic and commercial activity, environmental aspects, social cohesion and gender equality. To achieve this, as stipulated by the legal framework, the report should detail the extent of programme implementation in terms of actions and funding provided, any deviations in programme implementation, the relationship between results obtained and objectives set, the impact of interventions on the built environment, on the deficit in social services and facilities, and on demographic, social and economic problems. As will be explained, the report is decisive in establishing the continuity of further joint actions involving the Generalitat and the city council in the locality.

The obligation to conduct an *overall evaluation of the programme results* has been met by the government through the development of various internal and external reports, with the latter commissioned to various university research centres, private consultants and other government bodies. This set of materials has been submitted to the Catalan parliament and its findings published in a collected volume.[21]

21 See Nel·lo (2009). This work also contains a file with the essential features and the budget for each of the 92 neighbourhood programmes included in the five call programmes. The volume includes the following evaluation reports: a) Impact of the Law on Neighbourhoods on both the activities and budgets of the Catalonia public administration, b) Social incidence of the Law's application, c) Programme impact on the building and renewal of housing, d) Effects on the transversal nature of policies and administrative

From this assessment derives a large number of useful conclusions designed to guide future programme development. Given its summary format, the study on the social impact of implementation (Cardona et al. 2009) is of particular importance. This fairly simple exercise consisted of surveying those districts already into their fourth programme year, with recourse to the same socio-economic statistical evaluation procedures carried out at the time of selection. The results contain certain limitations since action effects are not always immediate and further analysis is needed, especially as many projects were still pending completion. However, the results themselves are very significant. As explained above, the system of parameters is the result of 16 statistical indicators accounting for demographic, urban, economic and social development. The higher the total score of each district, the more serious the problem. Now, as shown in Table 5.4, in the fourth programme year, the average score for first-call neighbourhoods dropped from 48.37 to 44.79, hence, not only did it not worsen, but it actually improved by 7.4 per cent overall. This improvement has been significant in six of the 13 districts where scores dropped by more than 10 per cent, in another three districts there was more moderate improvement, a drop of between 0 and 10 per cent. Finally, four neighbourhoods have continued to see their problems worsen, despite programme implementation, and witnessed an average 6.6 per cent increase in their scores.

This objective improvement was confirmed by a survey of citizen perceptions (Cardona et al. 2009): in 11 of the 13 first-call districts, the number of citizens who considered that life had improved in recent years is higher than those who believed otherwise. Furthermore, in each of the 13 districts, a large majority of citizens (85.7 per cent on average) stated that, even if in a position to do so, they would not live elsewhere.

However, this positive conclusion must be tempered in view of the results obtained by repeating the same evaluation exercise in the districts benefiting from the second call of the programme a year later, i.e. in 2009. At this time, when the effects of the crisis that began in 2007 could already be clearly perceived, the results were much less encouraging. So after four years of implementation, the situation measured by statistical indicators mentioned above, if not worse, had not improved noticeably. One could argue, of course, that without the intervention of the programme living conditions in these neighbourhoods would surely be worse. But in any case, this evolution shows very clearly the limitations of urban rehabilitation policies against the evolution of other more general economic and social variables.

organisation, e) International profile of the programme, f) Evaluation of induced public investment, g) Programme relationship with other urban renewal processes, h) Cooperation and dissemination of experience mechanisms.

Adapting to Changes

The tenth lesson to be drawn from the experience of districts in the Catalan neighbourhood programme is the need to maintain a constantly critical attitude towards the intervention instruments adopted. Indeed, analysis results in a reasonably positive general assessment for the first years of implementation. Nevertheless, after this first period a clear need to refine and adapt to economic, social and political changes emerged.

The first of these changes, and certainly the most decisive, was the alteration in the prevailing economic conditions. Whilst the first three years of operation saw the programme implemented within a context of Catalan economic expansion, as of 2008 the world entered a phase of instability and obvious difficulties. In Spain, the economic crisis has dramatically changed real estate market conditions and migration flows. The effects this may have on the accentuation of urban segregation processes in the long run are still uncertain. However, it is clear that in programme neighbourhoods, and others in similar situations, social issues – such as unemployment and the risk of exclusion – are already significantly apparent. This necessarily requires increased cohesion policies.

The second change was of a political and administrative character. As stated, with the fifth call (corresponding to 2008) the programme reached 92 neighbourhoods, 1,000 M€ of allocated investment, and benefits a population representing more than 10 per cent of Catalonia. The regional government commitment was to reach 100 neighbourhoods, a limit over which the programme could become denaturalised.

Based on these findings, and after examining the evaluation exercise outlined above, the government proposed the introduction of certain modifications to programme implementation, as approved by the parliament of Catalonia in the 2009 budget legislation. The amendments had the following objectives:

- Enhancing the performance of the Law on Neighbourhoods as a tool for the development of social cohesion policies on a territorial basis and transversal in nature, especially necessary in a recessionary period.
- Achieving the number of 100 districts benefiting from special attention under the programme and, after reaching this number, establish limit measures to preserve scopes of action consistent with the specificity of the instrument.
- Maintaining cooperation and aid in neighbourhoods that have already implemented all the actions foreseen in their respective programmes.
- Facilitating programme access to smaller municipalities experiencing significant difficulties in meeting programme contribution costs.
- Maintaining, despite budgetary constraints, the annual financial allocations to the Neighbourhood Management Fund and complement them with other funds, such as Local Government, Health, Employment, and others.
- Deepening innovations achieved in the administrative structure and in

> the working methodologies enabling administrative cooperation between the Generaltitat and municipalities, the transversal nature of policy implementation, and the involvement of neighbourhood organisations and associations.

Consequently, following the 2009 call, when the neighbourhood programme entered its second four-year period of implementation, changes have been made to allow for the existence of three different approaches to realities on the ground: comprehensive intervention projects, neighbourhood contracts, and the specific programme area for smaller municipalities.

Regarding the *comprehensive intervention projects*, the target of 100 districts participating in renewal processes was achieved in 2009. Henceforth, new localities will only be added depending on the number of those completing their respective programmes, i.e. each call accepts the same number onto the programme as neighbourhoods terminate their works. The idea is to maintain the consistency and specificity of the programme and the administrative ability to monitor and evaluate the ongoing projects with every possible guarantee.

Likewise, neighbourhoods completing the measures planned by their respective projects became eligible for a new scheme called a *neighbourhood contract*. The objective is to retain special collaborative mechanisms in those neighbourhoods that, despite the programme work completed, continue to require special attention from the authorities. As seen before, the Neighbourhood Programme is, in some respects, a type of shock intervention, which seeks, through the temporary concentration of procedures, to reverse the dynamics of degradation. However, the process of profound neighbourhood renewal is usually a task for more than a generation. Furthermore, it has not seemed advisable to abruptly discontinue the investment, let alone when special needs exist in a period of crisis. The contracts bind the Administration and the Generalitat as a whole with each municipality and provide local Development Programme funding. As noted above, actions to be financed are to be decided upon by a final evaluation report from the respective council, which must submit it to the Evaluation and Monitoring Committee on completion of all project procedures.

Finally, a specific sub-programme envisaged to promote comprehensive rehabilitation projects in municipalities with under 10,000 inhabitants was created. The objective of this mechanism, known as *Programa de villas con proyectos*, is to facilitate the incorporation of smaller municipalities as programme beneficiaries, resolving the main stumbling block they have faced so far: the inability to meet the 50 per cent funding contribution requirement. In fact, the law stipulated that government financing of projects could range from between 50 per cent and 75 per cent of the total so that, as from the 2009 call, it was established that for municipalities with under 10,000 residents regional government funding would account for 75 per cent of the total. Under the auspices of this alteration, 29 small and medium municipalities have already been able to proceed with project programmes.

The need to continue innovating and to adapt to changes in order to defend social cohesion in all towns and cities in Catalonia is, therefore, the tenth and final lesson from the experience of the Law on Neighbourhoods, with all its strengths and limitations.

References

Albors, J. 2008. La millora urbana des dels barris: marc instrumental, intervenció integral i oportunitats, in *Ciutats en (Re)construcció: Necessitats Socials i Millora de Barris*, edited by J.M. Llop, X. Valls, J. Albors and D. Mongil. Barcelona: Diputació de Barcelona, 259–68.

Amill, J., Berga, C. and Cascante, G. 2009. La transversalitat en el model de gestió del programa d'intervenció integral als barris, in *La Llei de Barris. Una Aposta Col·lectiva per la Cohesió Social*, edited by O. Nel·lo. Barcelona: Generalitat de Catalunya, 116–30.

Atkinson, R. 2003. Addressing urban social exclusion through community involvement in urban regeneration, in *Urban Renaissence? New Labour, Community and Urban Policy*, edited by R. Imre and M. Raco. Bristol: Policy Press, 101–20.

Borja, J. 1986. *Por unos Municipios Democráticos. Diez Años de Reflexión Política y Movimiento Ciudadano*. Madrid: Instituto de Estudios de Administración Local.

Cardona, Á., Homs, O. and Gay, J.. 2009. La incidència social de la Llei de Barris, in *La Llei de Barris. Una Aposta Col·lectiva per la Cohesió Social*, edited by O. Nel·lo. Barcelona: Generalitat de Catalunya, 60–74.

Castells, M. 1983. *The City and the Grassroots. A Cross-cultural Theory of Urban Social Movements*. London: Edward Arnold.

Cremaschi, M. 2005. *L'Europa delle Città. Accessibilità, Partnership, Policentrismo nelle Politiche Comunitarie per il Territorio*. Florence: Alinea.

De Gregorio, S. 2010. El desarrollo de las iniciativas comunitarias Urban y Urban II en las periferias degradadas de las ciudades españolas. Una contribución a la práctica de la regeneración urbana en España. *Ciudades*, 13, 39–59.

Domingo, M. and Bonet, M.R. 1998. *Barcelona i els Movimients Socials Urbans*. Barcelona: Fundació Jaume Bofill.

Donat, C. 2010. L'habitatge a la regió metropolitana de Barcelona, 1995–2006. *Papers. Regió Metropoliana de Barcelona*, 51, 44–60.

García-Almirall, P., Fullaondo, A. and Frizzera, A. 2008. Inmigración y espacio socio-residencial en la región metropolitana de Barcelona. *Ciudad y Territorio. Estudios Territoriales*, 157, 727–42.

García-Ferrando, L. 2008. Retos para un nuevo modelo de intervención en barrios: la *Llei de barris* de cataluña (2004). Cambios en las políticas de regeneración urbana. *Scripta Nova. Revista Electrónica de Geografía y Ciencias Sociales*, 270.

Giner, S. ed. 2002. *Enquesta de la Regió Metropolitana de Barcelona. Condicions de vida i Hàbits de la Població. Informe General*. Barcelona: Mancomunitat de Municipis de l'Àrea Metropolitana de Barcelona.

Gutiérrez, A. 2008. El mètode URBAN i la seva difusió com a valor afegit de la iniciativa comunitària, in *Ciutats en (Re)construcció: Necessitats Socials i Millora de Barris*, edited by J.M. Llop, X. Valls, J. Albors and D. Mongil. Barcelona: Diputació de Barcelona, 303–25.

Gutiérrez, A. 2009. *La Unió Europea i la Regeneració de Barris amb Dificultats. L'Acció de la Iniciativa Comunitària URBAN i la Construcció d'una Política Urbana Comunitària*. Lleida: Universitat de Lleida.

Harvey, D. 1973. *Social Justice and the City*. London: Edward Arnold.

Indovina, F. 1992. La città possibile, in *La Città di Fine Millennio*, edited by F. Indovina. Milan: FrancoAngeli, 11–76.

Leal, J. and Domínguez, M. 2008. Transformaciones económicas y segregación social en Madrid. *Ciudad y Territorio. Estudios Territoriales*, 157, 703–25.

López, J. and Rey, A. 2008. *Inmigración y Segregación en las Áreas Metropolitanas Españolas: la Distribución Territorial de la Población no Europea en el Territorio (2001–2006)*. Barcelona: Institut d'Estudis Territorials.

Martí-Costa, M. and Parés, M. 2009. *Llei de barris: cap una política de regeneració urbana participada i integral?* Barcelona: Generalitat de Catalunya.

Mier, M.J. and Botey, R. 2009. Quatre anys d'aplicació de la Llei de Barris, in *La Llei de Barris. Una Aposta Col·lectiva per la Cohesió Social*, edited by O. Nel·lo. Barcelona: Generalitat de Catalunya, 33–58.

Mongil, D. 2010. Intervención integral en barrios: conceptos, instrumentos y elementos de mejora. *Ciudades*, 13, 139–61.

Mucchielli, L. 2007. Les émeutes de novembre 2005: les raisons de la colère, in *Quand les Banlieues Brûlent. Retour sur les Émeutes de Novembre 2005*, edited by L. Muchielli and V. Le Goaziou. Paris: La Découverte, 11–35.

Muñoz, F. 2004. *UrBANALització. La Producció Residencial de Baixa Densitat a la Província de Barcelona*. Bellaterra: Universitat Autònoma de Barcelona, PhD, 3 vols.

Muñoz, F. 2006. Fer ciutat, construir territori. La Llei de Barris. *Activitat Parlamentària*, 8–9, 59–75.

Muñoz, F., Porcel, O., Pybus, M. and Valdelvira, M. 2009. La Llei de Barris i la revitalització urbana. Els efectes sobre la construcció i la rehabilitació d'habitatges, in *La Llei de Barris. Una Aposta Col·lectiva per la Cohesió Social*, edited by O. Nel·lo. Barcelona: Generalitat de Catalunya, 83–104.

Nadal, J. 2007. *Compareixença davant la Comissió de Política Territorial del Parlament de Catalunya*. Barcelona: Generalitat de Catalunya.

Nel·lo, O. 2004. ¿Cambio de siglo, cambio de ciclo? Las grandes ciudades españolas en el umbral del siglo XXI. *Ciudad y Territorio. Estudios Territoriales*, 141–2, 523–42.

Nel·lo, O. 2005. La nuova politica territoriale della Catalogna. *Archivio di Studi Urbani e Regionali*, 83, 39–70.

Nel·lo, O. 2008. Contra la segregación urbana y por la cohesión social. La Ley de Barrios de Catalunya. *Cidades. Comunidades e Territorios*, 17, 33–46.

Nel·lo, O. ed. 2009. *La Llei de Barris. Una Aposta Col·lectiva per la Cohesió Social*. Barcelona: Generalitat de Catalunya.

Nel·lo, O. 2011. The five challenges of urban renewal. The Catalan experience. *Urban Research and Practice*, IV(3), 308–25.

Nel•lo, O. 2012. *Ordenar el territorio. La experiencia de Barcelona y Cataluña.* Valencia: Tirant lo Blanch.

Parkinson, M. 1998. *Combating Social Exclusion. Lessons from Area-Based Programmes in Europe*. Bristol: Policy Press.

Pérez, V. and Sánchez, P. eds. 2008. *Memoria Ciudadana y Movimiento Vecinal. Madrid, 1969–2008*. Madrid: Libros de la Catarata.

Picorelli, P. and Saborit, V. 2009. La inversió pública induïda per l'aplicació del programa de barris, in *La Llei de Barris. Una Aposta Col·lectiva per la Cohesió Social*, edited by O. Nel·lo. Barcelona: Generalitat de Catalunya, 76–82.

Secretaria d'Habitatge. 2009. *Informe sobre el Sector de l'Habitatge a Catalunya 2008*. Barcelona: Generalitat de Catalunya-Departament de Medi Ambient i Habitatge.

Smith, N. 2006. Gentrification generalized: from local anomaly to urban 'regeneration' as global urban strategy, in *Frontiers of Capital. Ethnographic Reflections on the New Economy*, edited by M. Fisher and G.S.Y. Downey. Durham, NC: Duke University Press, 191–208.

Subirats, M. 2012. *Barcelona: de la Necessitat a la Llibertat. Les Classes Socials al Tombant del Segle XXI*. Barcelona: l'Avenç.

Trilla, C. 2002. *Preu de l'Habitatge i Segregació Social a Barcelona*. Barcelona: Patronat Municipal d'Habitatge.

Van den Berg, L., Braun, E. and Van der Meer, J. eds. 2007. *National Policy Responses to Urban Challenges in Europe*. Aldershot: Ashgate.

Ysa, T. 2011. *Programa de Barris 2004–2010: els Impactes de la Gestió en Xarxa*. Barcelona: ESADE.

Chapter 6
Cities and Urban and Metropolitan Regions in Spain: A New Agenda in a Global Context

Joan Romero-González[1] and Joaquín Farinós-Dasí[2]

The Return to Importance of Cities, Metropolitan Areas and Urban Regions as Relevant (Geo)Political Players

During the last two dramatic decades, the world has become flatter yet more spiky. This has been brought about by the political changes that led to the end of the Cold War, and to an even greater extent, by the tremendous changes generated by information and communications technology (ICT). Both Friedman (2006) and Florida (2008) have a point. Time and space have radically changed their traditional meaning. Over this period, the limitations and capabilities of the nation-state that emerged in Westphalia have also been changed, undergoing a remarkable transformation. Other legitimate and illegitimate powers have emerged or been consolidated during this time and other players move with greater ease in this globalised and interdependent world in which states (governments) are having difficulty adapting.

The economy, and in particular the financial system, has long been thinking globally, whereas politics has more difficulties in doing so and still prefers to think at the state level. Hence, people talk quite fairly about Economics 2.0 (even 3.0) and Politics 1.0. To avoid this, new forms of cooperation are explored (G-7, G-8, G-20, BRIC, *Next Eleven*, etc.) yet these are not always effective. Sometimes, as is the case in Europe, original forms of union are launched. These may be monetary or economic but are hardly likely to be fiscal or political. In other cases, various new and sometimes effective forms of regional cooperation are promoted, whether in Asia, Latin America or Africa. However, no one today disputes that in less than two decades the state has shed its traditional role as a result of the emergence of a truly new phenomenon whose consequences are as unknown as they are unpredictable: it is nothing other than a new division of power.

---

1 Universitat de València; Instituto Interuniversitario de Desarrollo Local; Departament de Geografia. Avenida Blasco Ibáñez, 28; 46010 Valencia (Spain). juan.romero@uv.es.

2 Universitat de València; Instituto Interuniversitario de Desarrollo Local; Departament de Geografia. Avenida Blasco Ibáñez, 28; 46010 Valencia (Spain). joaquin.farinos@uv.es.

The paradox in this process, marked by uncertainty, complexity, conflict and the speed of change, is that as the modern state has been transformed and seen its traditional outlines blurred, so cities have gained fresh prominence as political players at all levels from the local to the global. It is not the first time that this has happened, as is well known. In other periods in European history, cities were indisputably predominant (Boucheron et al. 2010: 311–25, Zeller 2010: 47–95) when states had not yet fully developed their potential.

The truth is that now cities (and not just the big global cities) have set up positions of strength on the geopolitical and economic chessboard and have tremendous opportunities to do well for themselves in this new reality; to occupy (or not) a prominent or unique place in this changing reality in which State borders are less relevant than they were 20 years ago. Urban regions, cities, and of course intermediate cities as well, compete and cooperate with each other (Pike et al. 2006). In the case of Spain, most of the time they have done so by relying on unbalanced growth models and using management approaches and methods imported from the private sector, driven by appropriable profits while forgetting their essential condition as public spaces for citizenship, community living and rights. Given this gradual process of dualisation, of 'breaking away', of wanting to get to the top table of globalisation at the cost of segregation and insecurity, there are many who on seeing the social consequences, aggravated by the recession, are now calling for greater citizen empowerment and new forms of urban governance to recover and strengthen the role of civil society, which is the real ground on which to build cities and metropolitan regions capable of maintaining an appropriate and sustainable balance.

In this new context, some metropolitan regions and cities are winners, and territories and others which are losers. Furthermore, this condition no longer depends on the capacity or size of their own states, but rather on the capacity and engagement of the public and private players concerned, their intelligence to agree on shared strategies, their readiness to build mobilising projects, their ability to pinpoint strengths based on a common purpose, to the point of renouncing the totem of global competitiveness and instead opting for a new development model or, more modestly, for an economy based on the provision of basic services in the general interest (economic, social, ecological, etc.) that ensure local people's quality of life is maintained. Territorial culture and intelligence are nowadays essential to do this (concreted as New Strategic Spatial Planning), and the appropriate scale for deploying their full potential is at the city and urban and metropolitan region level. In fact, the spiky world reflected in Florida's famous cartography is in reality a world of cities and urban and metropolitan regions, not states, because that is where things happen and where power, resources, innovation and talent are concentrated.

Hence, there is no better laboratory than cities and metropolitan areas in which to find out about the true extent of change; to examine the real scope of the transformations of a globalised economy and their impact on local labour markets; to assess the full extent of the increasing process of segmentation, insecurity and rising inequality in our societies and analyse its political, social and cultural consequences; to learn how to develop some sort of alternative for largely urban populations who are expressing a degree of insecurity, vulnerability, uncertainty and confusion about the future as never before in recent decades; to anticipate and address the root causes that explain growing urban unrest; to remedy the feelings of withdrawal, now widespread in European society, which hinder the recognition of the Other and make it extremely difficult to build multicultural societies which we will have to move towards in any case if only for strictly demographic reasons if we truly wish to maintain the essence of our socioeconomic model; in short, to imagine a new agenda, a new generation of public policies geared to a geopolitical, economic, social, cultural and environmental context that is very different and largely new.

These processes should interest us both as concerned citizens and as a political community, because today more than ever it is essential to give full meaning to the term sustainable development, especially in cities and urban regions. In other words, we must ensure an equilibrium between three cornerstones – economic competitiveness, social cohesion and prudent management of resources – whilst avoiding marked asymmetries between them so as not to upset the balance of the whole. Yet we also have to build in strategic approaches at an appropriate level, which now has to be the real city, that is to say the one that integrates an urban or metropolitan region defined by daily networks and flows between multiple cores, in which the limits between urban and rural areas disappeared long ago since, as Soja would put it, now even the rural is urban (Benach and Albet 2010), or at least it is no longer rural.

Metropolitan areas have become the main benchmarks in globalised capitalism. Regions really do matter (Henderson 2010). Nonetheless, this economic prominence has not been endorsed by a greater political role, an area where they remain weak, and for some (such as Lefèvre 2010) this means they are nothing but chimeras. Yet the fact cannot be ignored that not only has there been discussion, both academic (e.g. Heilnet and Kübler 2005, Phares 2004, Stephens and Wikstrom 2000) and in terms of decision-making (including Net-TOPIC URBACT 2010, Sansom 2009, 2011, Sellers and Hoffmann-Martinot 2008), but also that there has been some progress towards the politicisation of metropolitan areas. These attempts at local government reform have in general taken one of two main directions: the well-known decentralised or municipality-based public choice model and the concentration model. In the latter case, they have been implemented in different ways: through forced municipal mergers (as in the United Kingdom), through concentration of service delivery from a rational and professionalised perspective, the so-called 'neoprogressive' American model (Sager 2004), or through voluntary association and cooperation between municipalities. As noted by Blatter (2003: 503) with reference to cross-border cooperation, but in a point which is also applicable to its metropolitan counterpart (whether cross-border – Pendall and Puentes 2008 – or not), in spite of the advances and new forms of de jure cooperation, 'the only way to achieve joint action has always been and still is through "agreement" or "consent". What has changed over the years, though, is the institutionalised approach to reaching "agreement". Blatter (2003: 515) typifies changes in institutionalised ways of reaching agreement through a matrix (or dispersion diagram) between two criteria: Tightly and Loosely Coupled Institutions (based on differences between organisations and networks) and Instrumental or Identity-Building (based on Gölher et al. 1994, institutional theory).

Given the forced disappearance of municipalities either by law or as a result of a supposed comprehensive, technocratic or salvation rationality, in both cases increasingly questioned in terms of principles and results (as being unbending and not very adaptable to change, including those taking place in metropolitan areas), it appears that the use of new flexible and adaptable forms of agreement and cooperation, the third of the ways set out in the previous paragraph, is the one that brings new opportunities. Thus, democratic governance at the metropolitan level (which necessarily involves citizen participation) would now be a new essential cornerstone on which its most important players can use collective strategies to build the principles of territorial cohesion and coherence into their policies and projects (Pascual and Goldás 2010, Romero 2009, Romero and Farinós 2011). These voluntary groupings define their own scope and fields of action.

Thus, metropolitan management connects with new governance practices applied to spatial planning, the finest field for the application of governance (Faludi 2002, Farinós 2005a), by harmonising and strengthening two important processes: a) formal recognition of the metropolitan area based on planning (plans, guidelines and strategies as appropriate and depending on the importance of the tradition of planning instruments, closely tied to town planning and land use in the Spanish case), and b) simultaneously vindicating the role of and the need for strategic spatial planning at this level (METREX 2004, Salet et al. 2003). We believe the best option is strategic spatial planning (Albrechts 2010), not just because it has more potential (in terms of its objectives and the way of achieving them through consensus) but also as being more versatile and adaptable to variable geometries or 'fuzzy boundaries' (Farinós 2001), an intrinsic condition of metropolitan and urban areas and regions that evolve or change over time, thus enabling a new form of 'soft planning' (Faludi 2010, Haughton et al. 2010).

New Urban Dynamics since the Late 1980s

The 'return' of the city as a political player has taken place in a post-industrial scenario which is very different from that of the 1980s. Since then, and in Spain too, urban dynamics have been dominated by powerful processes of dispersion, decentralisation and segregation. Alongside the recentralisation of certain activities associated with advanced services and gentrification, centrifugal processes have prevailed, albeit selective ones. What has been defined as the explosion of the city (Font 2007) has taken place. Hence, the continued references to the real city, the sprawling city, dispersed city, edge city, city of cities (Nel·lo 2001), flexible city (Poli 2009), etc. as opposed to central city. They all boil down to residential dispersion or urban sprawl and increasing demands for land in peri-urban areas for industrial and tertiary uses and infrastructure (Feria and Albertos 2010, Seixas and Albet 2010). These are all very visible changes indicating new interdependencies and greater complexity and functional specialisation in urban and metropolitan regions. Yet what is more important than these territorial changes are their consequences for the social fabric, and here the evidence, again also in Spain, indicates that urban segregation and social inequalities have increased, especially since 2008, leading to a complex and uncertain social and economic scenario which features all the political consequences entailed by socially unsustainable drift.

The factors behind these processes are well studied. ICT has led to substantial changes in production processes, with the most significant being diffuse specialisation and the transformation of the value added chain. Social and cultural changes have introduced new patterns and new demands that have built relevant changes into the daily life of broad swathes of society. This has been possible because the evolution of income has enabled some social sectors to leave the centre and settle on the outskirts. Finally, a necessary condition for this to have taken place has been the spread across the territory of services and facilities, or at least the provision of infrastructure that gives access to supply centres, thereby increasing mobility.

Spain was one of the developed countries to opt for residential construction as the driving force of its economy. The period from 1997 to 2007 has been described as the most intense investment cycle in the country's history, well above the EU average and similar to investment levels in Asian countries. Yet it was unbalanced growth inasmuch as it concentrated much of its investment in residential and non-residential real estate, although there was also a degree of investment in production and more in human capital (Pérez 2011: 252–9). This investment, which was highly concentrated in metropolitan regions and coastal areas, led to the creation of 8.1 million new jobs, of which around 20 per cent were directly in the construction industry and over 50 per cent in low-productivity and low-skilled activities. We know what the root causes of this unique and intense process that led to a speculative property bubble were (Romero 2010). It was also accompanied by significant legislative changes that favoured it and aspects of the unique decentralisation of political power that began with the development of the regional state (*estado autonómico*), the subsequent process of creating new networks of players and the consolidation of new local and regional institutional contexts that would in many cases prove to be decisive.

The consequences of these processes are also well known. Firstly, we have witnessed an exacerbation of forced and voluntary daily mobility, with highly visible consequences in the shape of traffic congestion, increased energy use, more accidents and higher environmental costs. Secondly, the use of land for residential purposes has increased disproportionately and irresponsibly with all its economic, social, environmental and political consequences and the higgledy-piggledy location of residential activities and uses. This territorial dystopia reveals a highly visible process of regional disparities and *urbanalization* (Muñoz 2008) and the unprecedented destruction of cultural landscapes, with the consequent loss of the potential and value of land. These dispersion processes, which are far removed from the traditional Spanish compact city model, also increase the use of resources and energy, are less efficient in this use and raise the cost of service provision and delivery.

Thirdly, the sudden collapse of the residential construction sector has led to an unprecedented rise in unemployment, urban poverty and social exclusion. More than two million jobs have been destroyed since the beginning of the recession, of which half were in construction. The rate of job destruction has been comparable to the speed with which they were created during the previous 10 years. The property crisis has had a greater impact on those regions in which the progress of urbanisation had been greater, more business activity and employment associated with residential construction had been generated and, consequently, which were more exposed to the consequences of a crisis in the industry: metropolitan regions and coastal areas. Spain, which had cut its unemployment rate to historic lows and had maintained economic growth rates above those in the rest of Western Europe, has experienced the sharpest fall since 2008 and will take longest to come out of the recession. It is once again at the top of the list of European Union countries with double the unemployment rate of eurozone countries and with over 20 per cent of its labour force out of work in 2011.

This decline affects the central city and its historic district as well as large parts of metropolitan industrial belts. It is here where social segregation, population aging, poor provision of infrastructure and social facilities and housing stock obsolescence become exacerbated, where social segregation and isolation in which the so-called 'fifth carriage' that is emerging in our affluent societies is more packed, where insecurity, growing inequality, which is particularly pronounced in Spanish metropolitan areas (OECD 2011), social exclusion, lack of prospects and the multicultural society with all its problems, tensions and fears of living with foreigners (Bauman 2006, Oliver-Frauca 2006) are best seen.

Fourthly, it has long been obvious that the traditional administrative map is no longer adequate in Spain. Segregation and dispersion do not observe administrative boundaries and traditional municipal limits are no longer sufficient, relevant or useful.

Governing Urban and Metropolitan Regions in Spain

For Spain's central government, 'metropolitan area' means the grouping of municipalities created by a regional government in areas of high urban concentration for the management of one or more common services. From an alternative spatial, demographic and economic planning standpoint, a metropolitan area is a 'polynuclear space that forms a single housing and labour market, which in turn reflects the increased scale of collective living space and the different spatial strategies of economic players' (Feria 2010).

The processes that have taken place in Spain are no different to those which have occurred in its neighbouring countries. Over the period 1978–2008, Serrano et al. (2010: 4, 110, 114–15) noted a progressive metropolitanisation based on the criteria used in their study (density, population size of more than 200,000 inhabitants, production diversification, daily functional integration and potential for integration with other areas) which increased from the 15 metropolitan areas defined in 1981 (with a total of 204 municipalities and 19,598,751 inhabitants) to 24 in 2008 (534 municipalities and 24,358,296 inhabitants). Spanish metropolitan areas increasingly bring together more production, facilities and people, cover a greater area and have a larger number of municipalities.

In 1978, the system of Spanish cities was classified into two international metropolitan areas (Madrid and Barcelona), six national-regional metropolitan areas (Valencia, Seville, Bilbao, Málaga, Las Palmas and Saragossa), and seven areas undergoing metropolitanisation (Palma de Majorca, Vigo-Pontevedra, Gijon-Oviedo-Avilés, Alicante, Murcia, La Coruña and Valladolid). Twelve other urban areas with potential for metropolitanisation were added by then (Granada, Córdoba, Cádiz, Jerez, Vitoria, Santa Cruz de Tenerife, Pamplona, Salamanca, San Sebastián, Burgos, Almería and León, although finall Cádiz and Jerez were considered together), all of which had a group of municipalities with a capital that had more than 150,000 inhabitants.

By 2008, the original 15 metropolitan areas had been joined by another nine (seven of the former ones – Vitoria, Salamanca, Burgos, Almería and León dropped out – along with another two with some potential, namely Tarragona-Reus and Castellón de la Plana, thus reinforcing the Mediterranean axis). Based on these figures, and in the light of potential expansion area forecasts for 2015, the study proposes two first-tier urban functional regions (Madrid-Guadalajara-Toledo and Barcelona-Girona-Tarragona), a second tier consisting of three essentially national (Valencia-Gandía-Castellón, Bilbao-San Sebastián and Alicante-Murcia) and two metropolitan (Seville and Málaga) areas, a third tier consisting of another four regional metropolitan areas (Oviedo-Gijón-Avilés, Saragossa, Cádiz-Jerez and Vigo-Pontevedra-Santiago), and finally a fourth tier featuring the remaining metropolitan areas which have less potential.

Using another definition (the commuting variable), Feria (2010) identifies a total of 46 metropolitan areas in Spain with a total of 1,234 municipalities, 27,456,832 inhabitants (about 60 per cent of the Spanish population) and 13,254,066 dwellings, which simply confirms metropolitan areas as the main scenario for recent urban development dynamics in Spain. Once again, Madrid and Barcelona are in the first tier as two cities with a continental reach and more than 10 million inhabitants between them. In the second tier are Valencia, Bilbao and Seville as national areas each with more than a million inhabitants. The separation between the lower tiers is not as clear with the author describing their distribution as continuous.

In spite of this clear reality (whatever the criteria used to identify it; as the two previous ones or that of accessibility – see Pueyo et al. 2009, 2012), the remarkable failure to develop territorial governance mechanisms tailored to the evolution of the 'real' city in Spain is one of the best examples of the gap between territorial rhetoric and actual public policy. The studies by Feria and Albertos (2010) and Rojas et al. (2005) are good examples. The Spanish experience to date has been so limited that it could be said that there has been a collective failure to develop new forms of governance and an unmitigated failure in the always more complex and debatable field of territorial government. The problems in driving joint projects, institutional disagreements and administrative fragmentation prevail over the still poor list of best practices. For this reason, spatial planning and management in metropolitan areas in Spain remains one of the most important geopolitical challenges (Romero 2009).

Except for four, now forgotten, attempts to define metropolitan areas from the administrative and legislative standpoint, whose function was the coordination and management of municipal and territorial planning (Bilbao 1946, Valencia 1949, Madrid 1963 and Barcelona 1974; Serrano et al. 2010), and the recent law setting up the Barcelona metropolitan area in July 2010 (Nel·lo 2011), so far in Spain mechanisms of good governance or territorial governance which can successfully address the obvious mismatches between administrative structures designed to solve nineteenth-century problems and the territorial dynamics of the second decade of the twenty-first century, and are able to match instruments for intervention with the scale and operation of these areas by adapting representation and decision-making to the life of citizens and economic players, have not been developed. These mismatches between territory and function prompt variable or flexible geometries (Farinós 2001, 2010) which can be used to reinterpret territoriality (Ramírez 1994) as well as the possibility of identifying new areas as a basis for new initiatives for governance (Farinós 2006) and spatial planning and management ('soft planning', Faludi 2010, Haughton et al. 2010) and not the other way round, as has happened up until now whereby plans have necessarily had to be adapted to administrative boundaries which limits their meaning and effectiveness.

Any option of thinking 'from a metropolitan standpoint' in basic legislation has been completely abandoned in Spain. In fact, even the timid reference to it in the Local Government Regulatory Act 7/1985 has been removed. This latter consideration is highly significant as it demonstrates that the lack of initiative of the central executive and the scarce political interest shown by regional governments is added to by the indifference of local authorities as political players who, when faced with such a relevant issue, should be more interested in having mechanisms in place that enable territorial coordination and coherence and facilitate cooperation at the supra-municipal level. However, a strong centralist tradition and a patronage interpretation of politics, defects that have been slavishly reproduced at the regional government level, remain a heavy burden.

We have hardly any examples of metropolitan cooperation initiatives that go beyond the joint management of some basic services and some planning proposals in some central cities and their immediate area of influence. However, in terms of spatial planning and management, it has been virtually impossible to overcome the traditional municipal-based model. Given its inadequacy for the established framework, so far it has not been feasible to draw up suitable indicators for the metropolitan level or indeed an appropriate legal framework (Márquez 2009: 37–47), since recent legislative initiatives for major cities do not resolve anything. Finally, there is an insufficient political tradition or culture of cooperation (Farinós 2010), and this is probably the biggest stumbling block to promoting democratic governance in a composite, federal-style state which needs highly developed multilevel mechanisms for coordination and cooperation.

The outcome is a profusion of initiatives (not without confusion) that under different names attempt to implement Zoning Plans, Strategic Plans, isolated initiatives or mobilising programmes and projects almost always at the municipal, and therefore wrong, level. In many cases, these initiatives do not get beyond the level of a document that adopts the new rhetoric of territorial governance, far removed from the daily political practice to be found in each municipal area. The example of the hitherto failed metropolitan *Territorial Plan for the Protection of the Farming Land around Valencia* is one of the best instances. Their inadequate scale is exacerbated by pressure from specific, clearly productivist contexts and the fact that spurious yet obvious interests (with as many 'visions' as municipalities) do not stretch beyond the four years that constitute the political cycle.

After a period of 20 years of attempts at and announcements of metropolitan strategic spatial planning, it is only worth examining and following the evolution of some unique initiatives such as *Bilbao River 2000* (Mas 2011) and the 3rd Barcelona Metropolitan Strategic Plan. However, metropolitan area spatial plans have been further developed which are more orthodox and closer to physical planning, but less ductile and adaptable, as would be desirable given the unstable boundaries of metropolitan areas. Yet even in this case, the situation is far from satisfactory: a very small number of plans approved (11) or initially approved (1) out of the 46 possible, plans concentrated in a few regions (Catalonia, the Basque Country, Andalusia, Castile and León, and Navarre), with no match between their scope and the group of municipalities that make up the metropolitan area, and with some front-rank areas (such as Madrid, in truth a metropolitan region rather than a metropolitan area) and regions (such as Galicia) having significant metropolitan processes without any plan whatsoever (see Map 6.1). To these 12 metropolitan spatial plans that have been approved, another 10 can be added that are currently being drawn up and are at very different stages ranging from preliminary steps to the start of approvals (Feria 2010).[3]

3 The 11 plans already approved are, by approval date (from 1999 to 2010): the Pamplona County Urban Regulations, the Granada Urban Agglomeration Spatial Plan, the Valladolid and its Surroundings Planning Guidelines, the Bay of Cádiz Spatial Plan (now under review after the Bay of Cádiz-Jerez was nullified by judicial decision), the Central Alava Partial Territorial Plan, the West Costa del Sol Spatial Plan, the Bilbao Metropolitan Partial Territorial Plan, the Seville Urban Agglomeration Spatial Plan (approved in 2009 but not consistent with its initial approach), the Málaga Urban Agglomeration Spatial Plan,

The Spanish case is a return trip. In the 1960s and 1970s, in the middle of the growth and consolidation of a city model that repaired breaches and compensated for inequalities, setting up forms of metropolitan government in delineated areas was chosen as a way to improve management and service delivery and gain territorial coherence. Years later, from 1978 to 2008, the goal would remain the same, to influence the trend model of growth and increasing GDP and competitiveness, while at the same time correcting its tendency towards territorial imbalance (greater territorial cohesion). However, Serrano et al. (2010) argue that this more territorialised and polycentric government has reduced the overall rate of economic growth in Spain that would have been achieved with investment in productive capital that was more geared towards the pursuit of greater productivity and concentrated in large functional urban regions and in the Ebro and Mediterranean axes.

Beyond the obvious diversity of organisational forms used, the truth is that the consolidation of metropolitan areas in the 1960s and 1970s left some examples of interest in the coordinated management of basic services, a process that would be largely truncated at the time when the regions began to become a reality. Hence, while 'traditional' forms of metropolitan government were tried out in other areas in the 1980s, in Spain the few organisations that might have enabled the authorities to create some kind of supra-municipal government in large metropolitan areas were dissolved. Regardless of the differences between some of them, for instance Madrid and Barcelona, the fact is that the far-reaching process of *devolving* political power to regional governments in need of legitimacy was not only not harnessed to create new areas of metropolitan government but it was also used to eliminate any initiatives in that direction. The priority given to the consolidation of new regional powers and the fact that the Constitution maintained the provinces as local government entities left no political space for the metropolitan level. This was the fundamental cause behind the dissolution of the existing metropolitan entities and the impossibility of others prospering. The reluctance that local authorities have always shown, as in other cases, was added to this necessary condition. In other words, Spain has 'skipped' a traditional stage of metropolitan government without the latter having been able to develop owing to a lack of 'institutional space' due mainly to the excessive zeal of the old local and provincial governments, and especially of the new regional governments and parliaments.

---

the Barcelona Metropolitan Region Partial Territorial Plan (2010) and the Girona Urban System Urban Development Master Plan. Meanwhile the Donostia-San Sebastián Partial Territorial Plan is the only one at the initial approval stage. The 10 plans being developed are the León Urban Area Planning Guidelines, the Salamanca Urban Area Planning Guidelines, the Alfoz de Burgos Planning Guidelines, the Almería Urban Agglomeration Spatial Plan, the Huelva Urban Agglomeration Spatial Plan, the Campo de Gibraltar Spatial Plan, the Alicante and Elche Metropolitan Environment Regional Action Plan, the Castellón Regional Action Plan, the Spatial Planning Regional Guidelines with specific Guidelines for Central Asturias and the Tenerife Metropolitan Area Regional Action Plan.

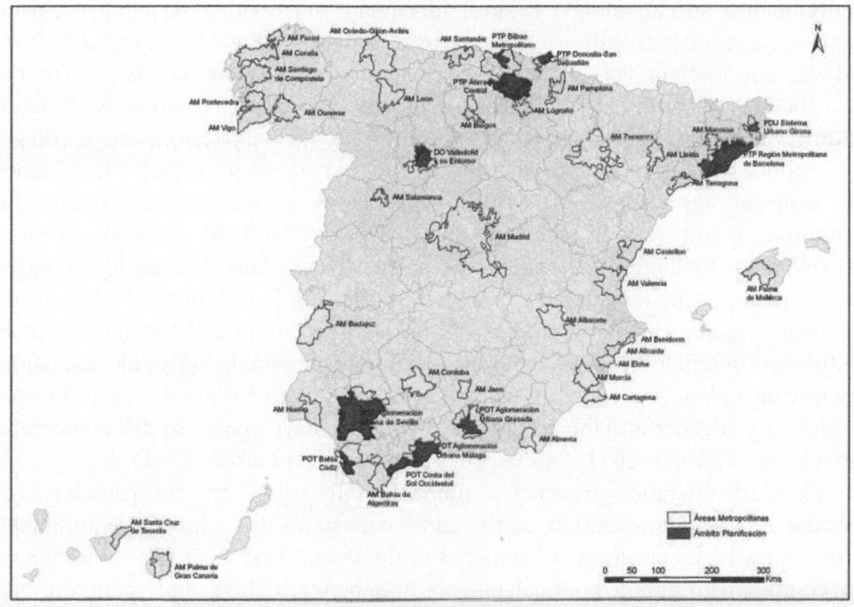

Map 6.1 Approved metropolitan spatial plans according to the 46 metropolitan areas in Spain

Source: Feria (2010, 2011).

But the facts are stubborn. The territorial dynamics and the social, economic and technological changes that have taken place since the 1990s, associated with an increasingly globalised context, have also demonstrated that in addition to a growing role for urban regions and places, as has been shown, there is also a strong need to agree on shared strategies at the appropriate level and among the players concerned. Furthermore, the most appropriate level for driving a new generation of public policies that entail substantial progress in territorial cohesion was not the municipal level, and nor were the democratically-elected government representatives the only players concerned, although they were very relevant. Cities and urban regions have grown even more as political players to the extent that if we were to get rid of maps showing the traditional municipal boundaries, we could draw up another, more realistic, cartography structured around networks of cities, urban regions and the corridors and flows between them.

Hence, once again, importance is given to the debate about the government of a 'real' or 'material' city with necessarily imprecise and changing limits ('fuzzy boundaries'), and to the focus on how to devise forms of coordination, cooperation and participation in areas in which capacity has increased and new opportunities and challenges have appeared that cannot be addressed by each of the players individually ('soft planning'). In short, this means overcoming the gap between an outdated administrative division that refuses to disappear and some real dynamics which, due to their very nature, do not abide by boundaries between, or the jurisdictions of, different levels of government and instead require new forms of multilevel and horizontal governance (Farinós 2005, Farinós and Romero 2008).

Political evolution and social change have today made possible the dominance of strategic approaches, of significant progress in the political culture of compromise and ways of understanding democracy with the extension of more participatory methods in decision-making processes. This has led to increased prominence for the idea that the best solution lies not in the successive extension of areas of management, or in the obsession with setting jurisdictional boundaries between different levels of government, or even in the establishment of 'cascading' hierarchical plans, but rather in the ability of the strategic players present in each region to go deeper into the principles of what we have agreed to call democratic governance (Farinós 2011, Pascual 2011, Romero and Farinós 2011).

Good democratic governance fundamentally rests on the principles of coordination and cooperation, and in composite states it is almost a requirement. Consequently, its progress is hampered to the extent that there is no willingness to compromise among political players and between them and economic and social players. This culture of compromise calls for organisation, instruments, institutional ability and an adequate regulatory framework to be mapped out. But it also requires as a necessary precondition the existence of a strong civil society. Only where the latter is dense enough has there been a greater chance of convergence and of building consensus between public and private players; agreements about strategic approaches to enhancing competitiveness, social cohesion and environmental protection in urban areas subject to the simultaneous challenge of competing in an increasingly globalised context, in a framework of restructuring the nation-state and impelled by the need to bring into being new values and principles based on sustainability.

The fundamental challenge for Spanish urban and metropolitan areas in this new era is reducing the gap between needs and new values that address them on the one hand, and institutional abilities on the other. The progress of values and approaches is significant. It is, however, more difficult to find appropriate ways to strengthen the abilities of the players and public authorities that facilitate progress in good governance or democratic governance. This is because these processes, as has already been noted, are political in nature and although ways and instruments to enhance democratic governance may be suggested in academic and professional circles, it is politics which has to drive solutions. Hence there are no interchangeable formulas, as these will depend on the context, political culture, democratic tradition and existing leadership capacity.

Furthermore, since decentralisation and deconcentration processes are asymmetric and constantly changing (flexible due to their own logic), the advisability of focusing more attention on the dynamics of urban regions rather than on traditional metropolitan areas, today overwhelmed by economic and social dynamics and territorial processes, is now being broached. At any event, it does not seem advisable to opt for the future establishment of new political and administrative figures of government (rigid and bounded by definition) at the metropolitan level to govern, promote, channel or reconcile changing processes whose definition is difficult if not impossible. In the current global context, territories need to find their place, their own fit. In a new stage, with a new generation of public policies (strategic initiatives, economic promotion, prudent land management, reducing social exclusion), the 'traditional' organisational solutions focused on more efficient provision of management of waste, transport, the water cycle and, where applicable, urban planning, have also become obsolete or insufficient (Farinós 2004). In short, we are faced with the problems of encapsulating the dysfunction between territory and function, and with the need to overcome the traditional approaches of managerial or bureaucratic government.

If not a new level of administration, territorial government in metropolitan areas does require the boost given by territorial plans that set binding guidelines which the municipal plans concerned (like any sectoral, regional or state plans that have a territorial impact) have to build into their planning. This is an option that in Spain is the responsibility of regional parliaments and governments but one which, in the light of experience (the Barcelona Metropolitan Territorial Plan is an exception), is unlikely to achieve the essential degree of consensus among the political players involved. In this case, it is highly likely that the exception of the Barcelona Metropolitan Territorial Plan and its development will remain the rule. This will depend on whether there can be new conditions which will undoubtedly be determined by: a) the depth of significant changes in the culture of cooperation, b) the existence of incentives that encourage or promote flexible and voluntary forms of cooperation between central and regional government and between regional and local government, c) the existence of political leadership on the part of the principal players, especially at regional level, to encourage and incentivise flexible and voluntary forms of multilevel cooperation and partnership, and d) the ability to put forward mobilising projects at a metropolitan level.

Some recent initiatives suggest that new directions to overcome imbalances that lead to opportunities being missed should at least be explored: a) greater institutional drive should be demanded so that metropolitan areas can occupy a better position in the context of a Europe of cities, b) the collective challenge of strengthening social and economic cohesion policies, thus avoiding the risk of social dualisation, should be undertaken and c) strategic approaches to involve the various players in the definition of objectives should be adopted, at least at the metropolitan level. In the Spanish case, turning towards new forms of governance in urban regions or metropolitan areas will require a review of the current basic legal framework, the adoption of new provisions at the regional level about levels of government and financing, and driving incentivised forms of cooperation in the style of those which have demonstrated their efficacy in France. The first two are complex issues that will certainly, as has happened before in Europe, face resistance and difficulties. As for the third of the proposed initiatives, some experiences suggest that there is enough leeway to enhance multilevel coordination mechanisms and incentivise cooperation methods.

As pointed out by Kübler and Heilnet, the probability of success in democratic metropolitan governance experiences is closely related to bringing together three factors: the strength of civil society, the functional legitimacy of local governments and the openness of the networks of relevant political players at different levels. As noted by the same authors, the European experience of best practice in democratic metropolitan governance indicates that it is dependent on favourable economic, social and political conditions, in short on favourable contexts. However, above all there are three crucial factors: a) the willingness to cooperate of the players involved, b) the existence of structures that incentivise cooperation, especially at higher institutional levels, and c) the ability for political leadership (Kübler and Heilnet 2005: 22–3). For very different reasons, none of the three crucial factors highlighted by these authors to explain the success of new forms of democratic metropolitan governance have been present in Spain.

The fact that when the regions were created they had a need for political legitimacy helps to explain the absence in Spain of sufficient political and institutional space to have developed 'traditional' forms of government at the metropolitan level in the past. It does not, however, justify it. The experience of other federal states such as Germany and Canada is a good example. In any case, there is nothing to stop new forms of democratic governance being developed in the future avoiding or at least being able to manage the risks of mimicking 'central' and 'regional' defects at this new level, and reproducing in it the same temptations for patronage (as also identified by Wassenhoven (2008) in reference to another Mediterranean country, i.e. Greece) and the risks of political 'parochialism'.

There is nothing to stop central and regional governments and parliaments from incentivising and promoting schemes for cooperation in this important field. Some examples indicate that sufficient leeway is available and progress can be made through the creation of basic consensus, voluntary cooperation and the establishment of incentive mechanisms by the public sphere. However, important opportunities to foster cooperation spaces and cultures at the local level have not been grasped. For example, the *National Local Investment Plan* pushed through in a hasty and improvised way by the central government in recent years has been an excellent missed opportunity to promote supra-local cooperation. Instead of starting from concepts based on a traditional municipal approach, it might have fostered municipal cooperation, in partnership with the regions, by incentivising the joint presentation of projects by a number of local governments and even by larger groupings of municipalities in the case of metropolitan areas.

Once more, the inertia of the sluggish traditional political culture has been insensitive to change and has had little ability and willingness to learn and develop new forms of interaction, management and governance. This has hindered devising other forms of partnership between levels of government and flexible and voluntary horizontal cooperation (between sector departments and territories) which have proven effective in other contexts where they have been used. They are probably the only ones that can be tried out in Spain today to drive consistent projects and overcome administrative fragmentation without reducing the number of divisions (municipalities) that are deeply embedded in the collective imagination and resist, despite their insufficiency, any attempt to introduce changes such as those that have taken place elsewhere.

New Agenda, New Policies and New Forms of Democratic Governance for Cities and Metropolitan Regions; Better Equipping the City and its Citizens

It is not too much to argue that these new urban dynamics require new approaches and new forms of government and democratic governance at local and metropolitan levels (Farinós 2006). There is already an extensive catalogue of best practice in dozens of cities and urban and metropolitan regions in other developed countries. Case studies in Western Europe show that there is no single model but rather a multiplicity of formulas conditioned by historical factors and specific contexts (Heilnet and Kübler 2005, Nel·lo 2010, Salet et al. 2003, Souto 2009), with notable examples of failed experiences and also very different degrees of 'success'. There is a lot of talent and very good ideas and experiences already available (Seixas and Albet 2010, Steytler 2009). Unfortunately, this is not the case with most Spanish cities and urban and metropolitan regions, where alongside few experiences of interest (Farinós et al. 2005) the level of cooperation between the players involved and the impetus behind innovative efforts are still insufficient.

Firstly, there are many good initiatives in the field of socio-institutional innovations and economic development policies that can enhance the quality of a territory and environment and strengthen the international dimension of cities. More and more importance is being given to the institutional context and high-quality integration between universities and technology institutes and the production system. Increasingly more attention is also being paid to strategic approaches. Great importance is always attached to clear public and private leadership in order to manage the changeover to a new production model that will have to be based on stimulating emerging activities and markets, on diversification (centred on local culture), and on the quality of services. Creative urban spaces will have to base this changeover on talent, tradition, technology and the ability to think together. The role of universities and the innovation and technology transfer system is essential. Also critical in the case of Mediterranean cities will be using a balanced model able to reconcile a tourism model founded on the quality and prudent management of resources and territorial quality. Territorial quality and good planning and management of cultural values and landscapes are one of Spain's strengths.

Secondly, there is a need to combat urban segregation and commit to better and more sustainable cities (Nel·lo 2011). This entails recovering the true meaning and balance between *polis*, *civitas* and *urbs* (in that order) and the same balance between the four cornerstones that guarantee the sustainability of a city and metropolitan region: a) the diversification of economic activity, b) the provision of necessary infrastructure and infostructure, c) wholeheartedly incorporating environmentally sustainable and healthy practices, and d) improving living conditions and the degree of social cohesion in a context where the underlying trends seem to be moving in another direction. In light of recent research (Tapada-Berteli and Arbaci 2011) this is perhaps the most difficult challenge: addressing segmentation and segregated reproduction processes, diversity management, caring for social risk groups, etc. But that is the issue that most identifies Europeans with the noblest sense of the city, citizens and citizenship – urban space and citizenship, in short – a society which is able to recognise the Other (Albet et al. 2006). Some good experiences show us that we have to begin with neighbourhoods, schools and social health services (Institut Montaigne 2009).

Thirdly, we have to pay due attention to degraded historic districts and metropolitan residential areas by fleshing out three basic principles – rehabilitation, regeneration and revitalisation – and demonstrate, as other cities in Europe and in Catalonia (with its Neighbourhoods Act) have done (Nel·lo 2011), that it is possible to combine culture, history, memory, productivity, competitiveness, modernity and social justice.

Fourthly, new forms of metropolitan governance, one of the main weaknesses in the case of Spain, must be developed. 'Thinking together' as a metropolitan region is crucial and therefore it is essential to enhance coordination and cooperation between political players and between public and private players. There are many and varied examples of good practice and imaginative forms of flexible cooperation, from Toronto and Vancouver to Melville in Australia, taking in various urban and metropolitan regions in Europe along the way. There are even good and very diverse and heterogeneous experiences in the intermediate cities group (Burgi 2009: 139–56, Pinson 2009, Young 2009: 109–12).

What are the risks in the Spanish case? Firstly in the field of politics; the resistance of local governments to taking part in forms of government or governance at the metropolitan level, the lack of clear leadership, of willingness to drive initiatives at the appropriate level, of the ability to manage networks, for building basic consensus, for devising new ways of coordination grounded in 'soft law' (Parejo 2007), are a major impediment. They are even more so if there are episodes of bad policy, of policy 'capture' or other institutional pathologies that seriously damage a city's reputation. Yet even though this resistance and these difficulties exist, especially when it comes to promoting forms of metropolitan government (Lefèvre 2010), it should not be inferred that there has been no progress, especially in the always more feasible field of developing forms of democratic metropolitan governance. Progress in new forms of government or governance in metropolitan areas has never been an easy process. Natural resistance at the local level, and in some cases from regional governments which have viewed the emergence of new forms of metropolitan government or cooperation with suspicion, has to be overcome. Any progress in new forms of government of the territory or metropolitan governance leads to political challenges for which solutions have to be devised and agreed that will vary depending on the specific context of each place.

A second risk is not addressing the rapid rise in inequality in cities and metropolitan regions and not adequately addressing urban discontent, its causes and its consequences (Romero 2011). The effects of offshoring are devastating for many production sectors and industrial districts, but the process is particularly intense and difficult to overcome in the metropolitan regions of Southern Europe, especially with the onset of the Great Recession. There is nothing new that Castells did not foresee even when the effects of globalisation were still not as evident (Castells 1998). He proposed an idea that with globalisation, work would become a global resource. However, labour markets, except for very specific segments, are local. In other words, in the 1990s the world achieved the unification of economic systems, companies could think globally, move and seek competitive advantage anywhere in the world, but people still remained tied to *places*, which in Europe basically means cities and urban regions.

There are many issues that need to be addressed, particularly in Mediterranean European countries. Can a European industrial region compete with a region in an emerging country or with other long established European ones? How can a region which has twenty-first century social and environmental rights compete with another that still works in social conditions comparable to late nineteenth-century Dickensian industrial Europe but with twenty-first century technology? Are traditionally advocated labour market reforms enough to improve the competitiveness of developed economies? Can the competitiveness of a Southern European industrial metropolitan region be improved through reduced wages, job insecurity and temporary work? Does the empirical evidence support this? How can meaning be given to the principle of 'equipping' people for the knowledge economy and entrepreneurship when education and training are not even a clear priority in countries like Spain? How can the shift from a low-skilled, labour-intensive production model to a knowledge-based system be managed? Where can decent employment be found for workers from industrial sectors in which only a few innovative companies survive? What are the working and life prospects of well-educated but poorly paid young people and the over 30 per cent of young people who have left school early? How far are we supposed to reach in the 'long-distance race'? What economic, social, political and electoral impact will these processes have in cities and urban regions?

The third risk consists of going down the tempting slope of unique projects. There are also examples of Spanish and European cities where there has been a 'festivalisation' or proliferation of unique projects which have then been reduced to what Sánchez Ferlosio has aptly defined as 'empty boxes'. These unique projects or events have sometimes been promoted to the detriment of urban policies for the entire city (Pinson 2009), thus increasing the perception among citizens that there is one city to be visited and another city, which is not as glamorous or even uniformly decent, to live in. If the imbalances are very marked, cleavage may occur between the various cities that exist in a city. But above all, there is a serious risk of confusing *civitas* with *urbs* (Capel 2004, 2009) when the order, as has already been noted, must always be the reverse: first *polis*, then *civitas* and finally *urbs*.

Finally, it is in cities and metropolitan regions where democratic innovations that go beyond traditional forms of representative or electoral democracy should be encouraged. Information, participation and public consultation are the inspiring principles, still in their initial phase, of this new generation of initiatives that are beginning to see the light in some European and American cities (Smith 2009).

Conclusions. Whither Change?

Firstly, forms and forums for coordination and cooperation at the metropolitan level have to be developed as an interstitial space for regional/sub-regional/local levels, which, to be efficient, also requires cooperation between central government and the regions. The current situation in Spain regarding democratic territorial governance is still highly uncertain. The academic and professional community has done its job and in general has been able to upgrade and fully take on board this new generation of approaches and proposals. But there has been no leap into the realm of policy. Politics has added the rhetoric of governance to its discourse, as it also did with sustainable development, but the practical effects are exceedingly meagre. Numerous experiences show that with existing levels of segmentation, fragmentation, overlapping and lack of coordination, divergence may end up imposing itself over convergence. Coordination and cooperation, which the Spanish Constitutional Court has been advocating in its rulings since the early 1980s, are the basic elements of good territorial governance, of an appropriate sense of state and commitment to managing the 'res publica'. But its paltry practical development is so far one of our greatest weaknesses. A short-term political culture and political strategy of polarisation (turned into the efficient instrument of the political 'business' of Spain's main parties) have prevailed over the culture of consensus. There have been too many documents and too little political will to agree on shared strategies.

The way of incentivising cooperation 'from the top down' has proved very positive and deserves much more attention than has been paid to it so far by the state and regional levels. The French example, despite all its caveats and shortcomings (Lefèvre 2010) shows that progress is possible. Thus, all that is required is to give follow through to these initiatives, even though the success rate is low and achievements are still modest. But there is no need to leave everything up to these hegemonic levels of governance (state), just as 'bottom up' mobilisation from the local level (municipalities) in order to promote voluntary groupings of them, with or without the support of other state levels, is also an option they have a right to exploit and develop. There are experiences including national and especially international cooperation networks (for instance METREX, Eurocities, etc.) as well as EU Community Initiatives and other institutions such as the Congress of Local and Regional Authorities of Europe (CLRAE) of the Council of Europe, the Committee of the Regions, the Council of European Municipalities and Regions (CEMR), the United Cities and Local Governments (UCLG), the Latin American

Forum of Local Government and the European Union, Latin American and Caribbean Local Government Forum.

It is imperative to better 'equip' metropolitan regions with the design of regional scenarios and commitments to cooperation between political players (Camagni and Trullén 2011). As in other traditionally federal countries (Steytler 2009), developing new forms of democratic governance in cities and urban regions is a priority if the intention is to make cities and metropolitan regions attractive places to concentrate and attract economic activity sustainably and as a guarantor of the right to an adequate quality of life for citizens. In accordance with the current Land Act, the principle of sustainability should also be fully fleshed out, especially at the local level, to drive strategic initiatives so that policies gain in coherence. Local governments legitimately aspire to gain financial security and jurisdictional clarity. This is another of the central points in the chapter on regional policy in Spain. It is at this level where the negative regional impact of public policies can most clearly be seen, given the lack of supra-municipal coordination guidelines to ensure greater territorial coherence. However, to achieve this, they will have to learn to look beyond their administrative boundaries and thrust aside localism.

Secondly, everything possible must be done to reverse the trend towards segregation and social division. The increase in social inequality and growing urban discontent must be a matter of priority concern in the immediate future. Academia also needs to focus greater attention on providing better information about the nature and extent of social change and its implications for urban and metropolitan regions in Spain. It is also necessary to know how to respond to the social consequences of the recession based on research that helps to understand the process of urban resilience, find new approaches to reverse job losses, particularly among vulnerable groups, and put a stop to the alarming increase in social problems in a context of budgetary constraints (URBACT 2010). This involves building a good grasp of social and cultural changes and the social fragmentation that hinders the construction of new stories and which have many implications for the future. Here we would highlight two: firstly, increased apathy, detachment and expressions of political cynicism, and secondly, support for new populist options. There are other possible scenarios and a contribution should be made to the academic debate by getting people, especially the younger generations, to understand that there is nothing that cannot experience setbacks or sudden change and nothing that cannot be changed, that there are no gains that cannot be lost, starting with those that best identify us and make life better for people: the welfare state and the democratic right to the city.

Thirdly, there must be pressure for more and better democracy from the local to the global level. Much remains to be done in fields that are now critical to reconcile politics with the people and democracy itself: participation, transparency, accountability and new consultation processes. The current rhetoric (governance, sustainable development, public information about plans and projects, etc.) as it has been 'customised' has made these processes and mechanisms as irrelevant as they are harmless. There is a need to give real substance to these concepts and drive other forms of change and innovation that are more radical, in the noblest sense of the term, and sometimes even more subversive.

In this respect, cities and urban and metropolitan regions are today privileged areas (in demographic, economic, social and cultural terms) in which to introduce democratic innovations at all levels: neighbourhood, district, city, metropolitan area, networks of cities, etc. Furthermore, there is already an acceptable bank of positive experiences of participation and cooperation, albeit naturally with resistors (Kübler and Tomàs 2010) and their difficulties and risks, as noted by Subra: tactical exploitation by politicians, the 'window effect' of associations and groups trying to make themselves heard outside the system of representation and radicalisation. We also know about the difficulties of determining which audience should be used for public debate and to reconcile relations between representative, participatory and direct democracy (Subra 2007). But where this argument is well made, it is not a 'farce' but rather quality democracy (Pascual 2011, Subirats 2006).

In the information age there is no excuse for the authorities, beginning at the local level, not to make transparent and accessible information available to citizens on the Internet, for example about the use made of their tax money, by creating virtual portals. There is no reason not to conduct extensive local consultation on relevant issues. There is no difficulty in carrying out Internet consultations. In fact, we already have relevant experiences in the field of democratic innovations and best practice in participatory democracy in many cities. There are also promising experiences in areas such as deliberative democracy, 'mini-publics', 'e-Transparency', 'e-Democracy' and 'e-Participation' (Smith 2009). Furthermore, there is also progress in good governance and democratic governance in cities and urban and metropolitan regions.

No less important is the temptation of elected officials in cities to evade their own accountability to the public and to elected representatives, of institutional avoidance, preferring to govern the city with pomp, by project, by creating 'advisory boards' and fictitious consensus between concerned players and stakeholders that award themselves the gloss and veneer of legitimacy for alleged results (which cannot be checked due to the opacity of accounts), of preferring the mobilisation of elites as 'city bosses' (Getimis 2010) at the price of the demobilisation of the masses, of propaganda (which also has to be paid for), of the marginalisation of lower socioeconomic groups (out of three parts of a population, one may be excluded with the support of the other two; once again 'divide and rule') and of the segmentation of urban public spaces (Pinson 2009). This is the route that some appear to want to take, but it is not the right one.

In short, Spanish cities and urban and metropolitan regions are the best test benches where environmentally-sustainable economic activity, social justice, democracy and new forms of government and governance – at a level (the metropolitan) which is critical to multilevel relations – have the crucial opportunity to improve substantially. This is where progress can best be made in public policy, where the 'de-constructed' state can be 're-structured', where politics has to become dignified, where parties change – at this stage profoundly – or should be replaced (by others or by other forms of direct citizen involvement, as happens in some cities in Australia) and where new forms of consultation and democratic participation that are being simultaneously demanded today in many parts of the world have to be developed.

References

Albet, A., Clua, A. and Díaz-Cortés, F. 2006. Resistencias urbanas y conflicto creativo: lo público como espacio de reconocimiento, in *Las Otras Geografías*, edited by J. Nogué and J. Romero. Valencia: Tirant Lo Blanch, 405–23.

Albrechts, L. 2010. More of the same is not enough! How could strategic spatial planning be instrumental in dealing with the challenges ahead? *Environment and Planning B: Planning and Design*, 37, 1115–27.

Bauman, Z. 2006. *Confianza y Temor en la Ciudad. Vivir con Extranjeros*. Barcelona: Arcàdia.

Benach, N. and Albet, A. 2010. *Edward W. Soja. La Perspectiva Postmoderna de un Geógrafo Radical*. Barcelona: Icaria.

Blatter, J. 2003. Beyond hierarchies and networks. Institutional logics and change in transboundary spaces. *Governance. An International Journal of Policy, Administrations, and Institutions*, 16(4), 503–26.

Boucheron, P., Menjot, D. and Boocne, M. 2010. *Historia de la Europa urbana. La ciudad medieval*. Valencia: Publicacions de la Universitat de València.

Burgi, M. 2009. Federal Republic of Germany, in *Local Government and Metropolitan Regions in Federal Systems*, edited by N. Steytler. Quebec: McGill-Queen's University Press, 137–65.

Camagni, R. and Trullén, J. eds. 2011. *Escenaris Territorials per a les Regions Europees: el cas de Barcelona*. Papers, 54. Barcelona: Institut d´Estudis Regionals i Metropolitans de Barcelona.

Capel, H. 2004. El futuro de las ciudades. Una propuesta de manifiesto. *Biblio 3W*, 551.

Capel, H. 2009. La Historia, la ciudad y el futuro. *Scripta Nova*, 307.

Castells, M. 1998. *La Era de la Información. Economía, Sociedad y Cultura*. Madrid: Alianza Editorial.

Faludi, A. 2002. Positioning European spatial planning. *European Planning Studies*, 10(7), 897–909.

Faludi, A. 2010. *Cohesion, Coherence, Cooperation: European Spatial Planning Coming of Age?* London: Routledge.

Farinós, J. 2001. Reformulación y necesidad de una nueva geografía regional flexible. *Boletín de la Asociación de Geógrafos Españoles*, 32, 53–71.

Farinós, J. 2004. Challenges of multi-level governance for spatial planning between local and regional levels. *Quaderns de Política Econòmica* [Online], 6, 81–95. Available at: www.uv.es/qpe/revista/ [accesed: 7 December 2011].

Farinós, J. 2005. Nuevas formas de gobernanza para el desarrollo sostenible del espacio relacional. *Ería*, *67*, 219–35.

Farinós, J. 2006. Gobernanza territorial de ámbito metropolitano, in *Los Procesos Metropolitanos: Materiales para una Aproximación Inicial*, edited by J.M. Feria. Seville: Centro de Estudios Andaluces, 155–65.

Farinós, J. 2010. *La Gobernanza en España. Realidad y perspectivas*, VI CIOT, Pamplona, Spain, 27–29 October 2010.

Farinós, J. 2011. Gobierno, buen gobierno, gobernanza y gobernabilidad de los territorios. Más de lo mismo no será suficiente, in *Dinámicas Territoriales, Políticas de Desarrollo Territorial Sostenible y Nueva Gobernanza Territorial en el Espacio Iberoamericano. Conceptos, Métodos y Tendencias*, edited by A. Olmos. Toluca: Latin American Territorial Observatory Network, 144–75.

Farinós, J. and Romero, J. 2008. La gobernanza como método para encarar los nuevos grandes retos territoriales y urbanos. *Boletín de la Asociación de Geógrafos Españoles*, 46, 5–9.

Farinós, J., Olcina, J. and Rico, A. 2005. Planes estratégicos territoriales de carácter supramunicipal. *Boletín de la Asociación de Geógrafos Españoles*, 39, 117–49.

Feria, J.M. 2010. La delimitación y organización espacial de las áreas metropolitanas españolas: una perspectiva desde la movilidad residencia-trabajo. *Ciudad y Territorio. Estudios Territoriales*, 164, 189–210.

Feria, J.M. 2011. La ordenación del territorio en las áreas metropolitanas españolas, in *Ordenación del Territorio y Urbanismo: Conflictos y Oportunidades*, edited by J.M. Jurado. Huelva: Universidad Internacional de Andalucía, 127–60.

Feria, J.M. and Albertos, J.M. 2010. *La Ciudad Metropolitana en España: Procesos Urbanos en los Inicios del Siglo XXI*. Madrid: Civitas and Thomson Reuters.

Florida, R. 2008. *Who's Your City?* New York: Basic Books.

Font, A. ed. 2007. *La Explosión de la Ciudad*. Madrid: Ministerio de la Vivienda.

Friedman, R. 2006. *La Tierra es Plana. Breve Historia del Mundo Globalizado del Siglo XXI*. Madrid: MR Ediciones.

Getimis, P. 2010. *Different Modes of Leadership and Community Involvement: Consequences for Soft Spaces*. ESF Exploratory Workshop on Planning for Soft Spaces across Europe, Delft, The Netherlands, 9–10 December.

Göhler, G. et al. 1994. Begriffliche und konzeptionelle Überlegungen zur Theorie politischer Institutionen, in *Die Eigenart der Instituionen: Zum Profil Politischer Institutionentheorie*, edited by G. Göhler. Baden-Baden: Nomos, 19–46.

Haughton, G., Allmendinger, P., Counsell, D. and Vigar, G. 2010. *The New Spatial Planning: Territorial Management with Soft Spaces and Fuzzy Boundaries*. London: Routledge.

Heilnet, H. and Kübler, D. 2005. *Metropolitan Governance. Capacity, Democracy and the Dynamics of Place*. London: Routledge.

Henderson, A. 2010. Why regions matter: sub-state polities in comparative perspective. *Regional and Federal Studies*, 20(4–5), 439–45.

Innerarity, D. 2009. *El Futuro y sus Enemigos*. Barcelona: Paidós.

Institut Montaigne. 2009. *Deghettoïsser les 'quartiers'*, June 2009. www.institutmontaigne.org

Kübler, D. and Heilnet, H. 2005. Metropolitan governance, democracy and the dynamics of place, in *Metropolitan Governance. Capacity, Democracy and the Dynamics of Place*, edited by H. Heilnet and D. Kübler. London: Routledge, 8–28.

Kübler, D. and Tomàs, M. 2010. Jeux d'échelles et démocratie métropolitaine. Leçons montréalaises. *Métropoles*, 7, 2–10.

Lefèvre, C. 2010. The improbable metropolis: decentralization, local democracy and metropolitan areas in the Western world. *Análise Social*, XLV(197), 623–37.

Márquez, G. 2009. Estructura de gobierno y administración del poder urbano en Galicia, in *Ordenación y Gobernanza de las Áreas Urbanas Gallegas*, edited by R. Rodríguez-González. Oleiros: Netbiblo, 31–132.

Mas, E. 2011. La revitalización del Área Metropolitana de Bilbao: la gestión de Bilbao Ría 2000. Aspectos territoriales. *Boletín de la Asociación de Geógrafos Españoles*, 55, 35–57.

METREX. 2004. *The Revised METREX Practice Benchmark of Effective Metropolitan Spatial Planning Practice*. Available at: www.eurometrex.org/Docs/Meetings/san_sebastian_2004/Benchmark_Dev_EN.pdf [accessed: 7 December 2011].

Muñoz, F. 2008. *Urbanalización. Paisajes Comunes, Lugares Globales*. Barcelona: Gustavo Gili.

Nel·lo, O. 2001. *Ciutat de Ciutats*. Barcelona: Empúries.

Nel·lo, O. 2010. Transformacions territorials a l'Àrea metropolitana de Barcelona. Una visió a partir de l'Enquesta de Condicions de Vida i Hàbits de la Població. *Papers. Regió Metropolitana de Barcelona*, 51, 1–117.

Nel·lo, O. 2011. The five challenges of urban rehabilitation. The Catalan experience. *Urban Research and Practice*, 4(3), 308–25.

Net-TOPIC URBACT Project. 2010. *Enhancing New Forms of Urban and Metropolitan Governance. Report of URBACT Project New Tools and Approaches for Managing Urban Transformation Processes in Intermediate Cities*. Available at: http://urbact.eu/fileadmin/Projects/Net_TOPIC/outputs_media/NeTTOPIC_T._Publication _3.pdf [accessed: 7 December 2011].

OECD. 2011. *Divided We Stand. Why Inequality Keeps Rising*. Paris: OECD.

Oliver-Frauca, L. 2006. La ciudad y el miedo, in *Las Otras Geografías*, edited by J. Nogué and J. Romero. Valencia: Tirant Lo Blanch, 369–88.

Parejo, T. 2007. Nuevas formas de derecho blando para la ordenación del territorio. Enseñanzas de la Estrategia Territorial Europea, in *Territorialidad y Buen Gobierno para el Desarrollo Sostenible. Nuevos Principios y Nuevas Políticas en el Espacio Europeo*, edited by J. Farinós and J. Romero. Valencia: Publicacions de la Universitat de València-IIDL, 119–45.

Pascual, J.M. 2011. *El Papel de la Ciudadanía en el Auge y Decadencia de las Ciudades*. Valencia: Tirant Lo Blanch.

Pascual, J.M. and Goldás, X. 2010. *El Buen Gobierno 2.0. La Gobernanza Democrática Territorial*. Valencia: Tirant Lo Blanch.

Pendall, R. and Puentes, R. 2008. Land-use regulations as territorial governance in U.S. metropolitan areas. *Boletín de la Asociación de Geógrafos Españoles*, 46, 181–206.

Pérez, F. 2011. La crisis en España: la herencia de un crecimiento desequilibrado, in *Pasado y Presente: de la Gran Depresión del Siglo XX a la Gran Recesión del Siglo XXI*, edited by P. Martín-Aceña. Madrid: Fundación BBVA, 252–82.

Phares, D. ed. 2004. *Metropolitan Governance without Government?* Brookfield: Ashgate.

Pike, A., Rodríguez-Pose, A. and Tomaney, J. 2006. *Local and Regional Development*. London: Routledge.

Pinson, G. 2009. *Gouverner la Ville par Projet. Urbanisme et Gouvernance des Villes Européennes*. Paris: Sciences Po.

Poli, C. 2009. *Città Flessibili*. Torino: Instar Libri.

Pueyo, A., Jover, J.A. and Zúñiga, M. 2012. Accessibility evaluation of the transportation network in Spain during the first decade of the 21st century, in *Territorial Implications of High Speed Rail*, edited by J.M. de-Ureña. Farnham: Ashgate, 83–104.

Pueyo, A., Calvo, J.L., Jover, J.M. and Zúñiga, M. 2009. Visualización de los procesos territoriales desde el análisis de la evolución de la población y de las infraestructuras, in *Cohesión e Inteligencia Territorial. Dinámicas y Procesos para una Mejor Planificación y Toma de Decisiones*, edited by J. Farinós, J. Romero and J. Salom. Valencia: Publicacions de la Universitat de València-IIDL, 183–214.

Ramírez, J.L. 1994. *Los límites de la democracia y la educación*. Lleida: Universitat de Lleida.

Rojas, E., Cuadrado-Roura, J.R. and Fernández-Güell, J.M. eds. 2005. *Gobernar las Metrópolis*. Washington, DC: Interamerican Development Bank.

Romero, J. 2009. *Geopolítica y Gobierno del Territorio en España*. Valencia: Tirant Lo Blanch.

Romero, J. 2010. Construcción residencial y gobierno del territorio en España. De la burbuja especulativa a la recesión. Causas y consecuencias. *Cuadernos Geográficos de la Universidad de Granada*, 47, 17–46.

Romero, J. 2011. Ciudadanía y democracia. El malestar urbano y la izquierda posible hoy en Europa. *Biblio 3W. Revista de Geografía y Ciencias Sociales*, XVI(932), 1–12.

Romero, J. and Farinós, J. 2011. Redescubriendo la gobernanza más allá del buen gobierno. Democracia como base, desarrollo territorial como resultado. *Boletín de la Asociación de Geógrafos Españoles*, 56, 295–319.

Sager, F. 2004. Metropolitan institutions and policy coordination: the integration of land use and transport policies in Swiss urban areas. *Governance: An International Journal of Policy, Administrations, and Institutions*, 18(2), 227–56.

Salet, W., Thornley, A. and Kreukels, A. eds. 2003. *Metropolitan Governance and Spatial Planning. Comparative Case Studies of European City-Regions*. London: Spon Press.

Sansom, G. 2009. *International Roundtable on Metropolitan Governance*. ACELG (Australian Centre of Excellence for Local Development). Sydney, Australia, 14–15 December.

Sansom, G. 2011. *Second International Roundtable on Metropolitan Governance*. ACELG (Australian Centre of Excellence for Local Development). Brisbane, Australia, 19–20 August.

Seixas, J. and Albet, A. 2010. Urban governance in the south of Europe. Introduction to the special issue. *Análise Social*, XLV(197), 611–20.

Sellers, J. and Hoffmann-Martinot, V. 2008. Metropolitan Governance, in *World Report on Decentralization and Local Democracy,* edited by United Cities and Local Governments. Washington, DC: The World Bank, 259–83.

Serrano, A. et al. 2010. *Informe de Prospectiva a partir de las Transformaciones Territoriales tras 30 años de Constitución Española.* Madrid: Dirección General para el Desarrollo Rural Sostenible.

Smith, G. 2009. *Democratic Innovations*. Cambridge: Cambridge University Press.

Souto, X.M. 2009. *Áreas Metropolitanas Galegas*. Santiago: Xunta de Galicia.

Stephens, G.R. and Wikstrom, N. 2000. *Metropolitan Government and Governance. Theoretical Perspectives, Empirical Analysis, and the Future*. Oxford: Oxford University Press.

Steytler, N. 2009. *Local Government and Metropolitan Regions in Federal Systems*. Quebec: McGill-Queen's University Press.

Subirats, J. 2006. La gobernabilidad de las políticas territoriales. Formulación participativa y gestión concertada, in *La Nueva Cultura del Territorio*, edited by A. Tarroja and R. Camagni. Barcelona: Diputació de Barcelona, 389–408.

Subra, P. 2007. *Géopolitique de l'Aménagement du Territoire*. Paris: Armand Colin.

Tapada-Berteli, T. and Arbaci, S. 2011. Proyectos de regeneracion urbana en Barcelona contra la segregacion sociospacial (1986–2009): ¿solución o mito? *ACE: Architecture, City and Environment*, 17, 187–222.

URBACT. 2010. *Urbact Cities Facing the Crisis. Impact and Responses*. European Union, European Regional Development Fund.

Wassenhoven, L. 2008. Territorial governance, participation, cooperation and partnership: a matter of national culture? *Boletín de la Asociación de Geógrafos Españoles*, 46, 53–76.

Young, R. 2009. Canada, in *Local Government and Metropolitan Regions in Federal Systems*, edited by N. Steyler. Quebec: McGill-Queen's University Press, 107–35.

Zeller, O. 2010. *Historia de la Europa urbana. La ciudad moderna.* Valencia: Publicacions de la Universitat de València.

Chapter 7
The Governance of French Towns: From the Centre–Periphery Scheme to Urban Regimes[1]

Gilles Pinson[2]

Introduction

This chapter defends and illustrates the idea according to which studying the way French towns and cities are governed through the mere centre–periphery relationships analytical framework is no longer appropriate. This approach, which has long prevailed in Europe and particularly in France, considers that the 'vertical' relationships between the state's central and field administrations, on the one hand, and the local political actors and local administrators, on the other, are the most important facets to take into consideration in order to understand the way localities are governed. Besides, this framework attributes a decisive importance to the regulation and policymaking functions of central state representatives and tends, in parallel, to neglect the 'horizontal' dynamics of mobilisation, alliance and conflict entwining local actors, whether political or socio-economic. This approach is intimately linked to the history of the construction of territorial states in Europe which was characterised by the progressive expansion of centres into initially rebellious peripheries (Elias 1975, Tilly and Blockmans 1994). The challenges of military and political control of these peripheries prevailed for such a long period of time that they justified both institutional arrangements intended to limit the power and responsibilities of local governments and interpretative schemes emphasising the influence of the central state representatives.

The recent evolution in the relationships between the state and local governments and in policymaking processes, along with the transformations in capitalism and its relationships with urban spaces and the evolution of urban societies require a reassessment of the way French cities governance is analysed. The attribution of an ever increasing number of functions to urban governments, the competition between cities triggered by globalised capitalism and state policies more concerned

1 An early version of this text was published in 2010 in *Análise Social. Revista do Instituto de Ciências Sociais da Universidade de Lisboa*, 197, 717–37.

2 Université de Lyon; Sciences Po Lyon. 14, Av. Berthelot; 69365 Lyon Cedex 07 (France). gilles.pinson@sciencespo-lyon.fr.

with competitiveness than with redistribution, have made French cities not just spaces for the implementation of public policies but also actors in the elaboration of the policies and the strategic visions inspiring them. Local actors (elected officials, civil servants, economic actors, social movements and associations) are no longer in a subordinate position as regards determining and implementing urban policies. The 'horizontal' relationships linking those local actors can be now considered as the first explanation of the content of urban policies and of the way they are conceived and implemented. Therefore, it seems legitimate to henceforth approach urban policy/policies in France through theoretical tools which attribute a primordial role to the horizontal interactions between urban actors, groups and organisations, as well as to the conflicts, alliances and coalition logics in which they are bound up as factors explaining the structure of the urban agenda and the content of urban policies. Among these theoretical tools, the 'urban regimes' approach developed in the United States by Clarence Stone (1989; cf. also Orr and Johnson 2008) and his followers seems particularly fruitful. The merits of this approach will be subsequently presented in this chapter, but we can already state that one of them is the fact that it consents to articulate a political economy approach, sensitive to the structuring character of the capitalist framework, with a political science approach attentive to the political and social logics which shape urban policy and policies, whilst also not overestimating the role played by the influence of upper government tiers.

In this chapter, I shall first consider the theoretical approaches that have dominated the social science literature on local government in France and present the empirical reality – establishing strong asymmetries in the relationships between the state and localities – in which these approaches were rooted. I shall seek to demonstrate how these approaches have left strong imprints on contemporary scientific production whereas the empirical substratum at their origin has been substantially modified. In the second part, I shall consider how the transformations which occurred before and after the 1982 decentralisation laws finally allowed alternative analytical frameworks to emerge. These approaches, and the governance approach in particular, have permitted the re-evaluation of the role played by urban actors in the governance of cities. In this section, I shall plead for an extension of these works through the deployment of urban regimes research programme and theory. Finally, in a third and last part, I shall try to widen the subject to places beyond France and examine the possibility of defining and qualifying Southern European urban regimes.

The Long Goodbye to the 'Local Political Administrative System'

The notion of a 'local political administrative system' is part of the theoretical frameworks that have long dominated the approach to local politics in France. Developed by the researchers of Michel Crozier's *Centre de Sociologie des Organisations* (CSO), this notion refers to an approach that conceives of local political life, urban politics and local actors only through the issue of their relationships of dependence and submission to central state representatives and their attempts to skirt the rules imposed by the centre. This kind of approach has profoundly marked the French way of dealing with cities and urban policies.

The Erosion of Local Autonomy

Nevertheless, we must recognise that the approach which favours the dependent relationship between 'local' and 'state-national' is rooted in an empirical and historical reality that has witnessed the continued erosion of local autonomy, in particular that of cities, as a central power asserted itself, firstly royal, then imperial and lastly republican. While it has undoubtedly been weaker than in Middle Europe (Italy, Rhineland Germany, Flanders), the communal movement nevertheless did touch France. In the 'historic interlude' (Weber 1982) that separates the fall of the Roman empire and the failure of the Carolingian attempts to re-establish a Christian West on the one hand, and the emergence of nation-states in the fourteenth century, certain French towns liberated themselves from feudal bondage and the power of the bishops and obtained franchises and privileges from the crown (Chédeville et al. 1980). An urban consular power quite close to the city-state model was established, in particular in areas of what today makes up southern France. However, this communal movement was not sufficiently robust to oppose the influence of the rural nobility and the precocious emergence of a centralised royal power, which from the sixteenth century onwards progressively corroded the city privileges and replaced consular powers with those of crown representatives. Royal absolutism then brought a definitive end to this historic interlude.

The revolutionary period, and later the imperial episode, proved unfavourable to urban powers and worsened, rather than corrected, the process of centralisation. The revolution created the communes which, while again taking up the structure of the Catholic Church parishes, founded the basis for the eternal problem of extreme communal fragmentation (41,000 at the time of the Revolution and still 35,568 today). In its rationalistic logic, this endowed all communes with the same status be they urban or rural. Furthermore, the urban power legitimately associated with the power of the mercantile corporations and, therefore, with intermediary bodies, was not reinstated by the revolutionary governments. The administrative power was transferred to the state representative, above all the prefects created in 1800. Neither the empire nor the restored nineteenth-century monarchies ran counter to this centralising tendency. On the contrary, even if the municipal elections suspended in 1797 are reinstated in the 1830s under the Restoration, the nineteenth century was characterised on the one hand by a strengthening of the central state and its military and civilian infrastructures and, on the other hand, by the unification of the public national space, thanks to the development of the press and parliamentarism, all factors that prove in no way favourable to any resurgence in urban power.

One has to wait for the stabilisation of the Republican regime from 1875 onwards to encounter a new historical window of opportunity more favourable to the re-emergence of urban powers. The founders of the new regime carried the explicit intention of reducing the central state apparatus, compromised by its participation in the imperial regime and, in order to achieve that, backed the construction of the Republic 'from the bottom up' and, in particular, from the commune level. An era then begun that some call the 'the town hall era' (*Politix* 2001). The communes obtained the freedom of action to innovate in matters of public policies. That period saw towns – not the central state – invent urban planning (Gaudin 1985) and trace the outline of the welfare state. This capacity for innovation was also exercised during the first decades of the twentieth century in the field of economic development, in particular in the communes directed by socialists willing to experiment with a French style 'municipal socialism'.

Nevertheless, that emancipative drive of the urban powers swiftly got broken by censorship of the *Conseil d'Etat*, the judicial body responsible for the control of the legality of government actions, and the return in force of the central state apparatus between the two wars. This movement towards recentralisation was confirmed by the Vichy Regime which 'nationalised' urban planning and makes the central state a key actor in urban and construction policies. However, it was particularly with the 4th Republic (1946–58) and even more especially so with the 5th Republic (1958–) that the process of centralisation and the reduction of local governments' room for manoeuvre attained their high water mark. Immediately after the war, an elite consensus emerged favourable to the construction of a powerful interventionist social state able to the rebuild the country and to orientate its economy. The 4th Republic laid the foundations with nationalisations, the foundation of the social security system and the creation of a powerful planning agency: the *Commissariat Général au Plan*. The 5th Republic under de Gaulle even reinforced the central state by systematising the powers of the state's upper echelons over an increasingly extended range of social activities and economic sectors (François 2008, Jobert and Muller 1987). This interventionist state, a kind of 'revival' of Colbertism, also took on urban and regional challenges and implements what Brenner (2004), in accordance with Martin (1989), qualified as 'spatial Keynesianism'. Among the main institutional developments of these political strategies is the creation of the *Délégation Interministériel à l'Aménagement du Territoire et à l'Action Régionale* (DATAR) in 1963, an agency meant to contain the growth of the Paris region and to orchestrate the industrialisation of the poorest regions and equip them with modern infrastructure. During this period, the central state also took control of most of the planning, housing and infrastructure policies in cities.

The gradual rise in power of an interventionist and planning state goes hand in hand with the stabilisation of a specific power system associating state representatives and local officials that sociologists from CSO termed the 'local political-administrative system' (système politico-administratif local – SPAL) (Grémion 1976, Worms 1966). This system was mainly characterised by its asymmetry given sharp resource inequalities between state representatives richly endowed with financial and technical means as well as the legitimacy for action, on the one hand, and the elected local authorities and local civil servants who were progressively deprived of such resources as state interventionism grew on the other.

That intervention has, in fact, deprived local actors, including those in towns, of their capacity for taking initiatives and innovating in matters of public policies. Their role was progressively limited to maintaining a local consensus and acting as intermediary between local society and state administrations. Even if some prominent local elected officials holding multiple offices at local and national levels were able to negotiate with the state representatives the way national policies were locally implemented, even if the associations of elected local entities and in particular the Association of the Mayors of France (Association des Maires de France – AMF) allowed mayors to influence the legislation concerning local communities (Le Lidec 2001), even if thanks to their presence in parliament – and in particular in the Senate – elected local authorities were able to block attempts at reforming the local institutional system, overall, the state central and field services had acquired a strong and almost monopolistic grip over urban matters. The central technocratic elites of the de Gaulle regime at the time considered the elected local officials as forces which resist change and modernisation and the *département* Prefects as their allies. Thus came the idea of promoting the regional level as the new scale for central state action and establishing powerful agencies such as DATAR capable of overcoming obstructions put in place by local elites.

Naturally, such developments did nothing to improve the status of 'local' matters as far as social sciences studies are concerned. Along with the sociologists of the CSO, some political scientists (Lagroye 1973) depicted the elected local officials as busier consolidating electoral strongholds and mobilising the consensus in local society around state policies rather than trying to innovate or elaborate projects for their territory. In the 1960s and the 1970s, French urban sociology, when it was not totally insensitive to the political dimension, portrayed local political actors and administrators as totally marginalised by a central power that was carrying out an in-depth reorganisation of urban spaces for the benefit of the major national industrial groups (Castells and Godard 1974).

Decentralisation and Survival of Analytical Models

Paradoxically, the decentralisation laws of 1982 and 1983 which considerably upset the country's institutional architecture and centre–periphery relations did not fundamentally transform the outlook of the social sciences on local and urban policies. Voted through at the beginning of the first presidential mandate of socialist François Mitterrand (1981–8), these laws carried, nevertheless, important changes: they put an end to prefect tutelage over the acts of local communities replacing control over opportunities by a mere judicial control over the conformity of acts; they transferred strategic competences from state services to the local governments (the communes inherited urban planning, transportation, social action, among other competences); they made the rules for matters such as staff recruitment and access to credit more flexible. However, one should not attribute more importance to this surge in decentralisation than it actually bore in reality. The law did, more often than not, simply legally validate an evolution that had already begun back in the 1960s and which witnessed a gradual strengthening on the part of certain local communities in their capacity to produce public policies. What Lorrain (1989, 1991) has qualified as a 'silent change', above all, involved large cities which, even before decentralisation, had held an autonomous capacity for innovation and action in sectors where state services had invested little: public transport, economic development, the rehabilitation of rundown housing stock, culture, social action, etc.

However, Lorrain's findings have remained relatively isolated in French literature concerning local policies. Despite decentralisation, most scholars remained obsessed with the social figure of the '*notable*', i.e. those archaic and omnipotent local elected officials able to monopolise relations with the centre. In the years following the reform, a certain number of studies drew stern conclusions regarding decentralisation (Mabileau 1991, Mény 1987, 1992, Rondin 1985): rather than a democratisation of local political spaces and a strengthening of innovation in local policies, decentralisation further increased the already considerable power of some *notables* – mayors, presidents of general and regional councils – who recycle the same logic of action that they already practised prior to centralisation consisting in favouring courting the centre rather than mobilising those forces constituting local societies to promote a local development project. According to those studies, the decentralisation reforms missed their target. Territorial fragmentation, the quite undemocratic character of decision-making systems and the weak mobilisation of local resources within the fabric of public policies remained, according to those studies, important traits of local political systems.

If this kind of analysis, which continues to prevail in French academic fields interested in local and town policies, in particular in the political science field, is certainly the reflection of an empiric reality characterised especially by the weight of executives in local institutions, it is also the product of a rooted way to focus on elected officials and institutions and to systematically underestimate the polycentric and pluralist character of urban societies when looking at urban policies that still prevails. Not only does the 'hold of the state curb the decentralisation of the outlook' as Cadiou indicates (2009: 35, as do Briquet and Sawicki 1989), but even when the urban policies and politics are taken into consideration, the majority of French political science studies systematically endow the elected entities, the political institutions, the logics of political competition and the building of electoral strongholds with a highly privileged role. It is very much the case with the studies carried out during the 1980s and the 1990s in accordance with the research by Lagroye (1973) on Jacques Chaban-Delmas, Mayor of Bordeaux from 1947 to 1995. One has to recognise that those studies (Garraud 1989, Petaux 1982) contain the huge merit of having updated the conditions under which the profession is exercised by the local elected officials and, in particular, that of the logic of the construction of 'eligibility' (Abélès 1989), of mobilising support and of constructing electoral strongholds. They have demonstrated that local spaces are actually political spaces. Nevertheless, one can reproach them over two biases. First, the approach to urban government is systematically centred on elected officials and on the 'politics' side of their activities. Whereas, with a few exceptions (Fontaine and Le Bart 1994), issues such as local administrations management, the construction of projects, and the activation and coordination of public policies are often neglected. This focus on issues, spaces and practices of electoral competition tends, therefore, to render urban societies invisible, to neglect the materiality of the city and the set of problems that urban political actors may be faced with when dealing with urban society and materiality. Finally, despite the clearly sociological orientation of the studies, local societies are not made more palpable when compared to previous works. They are reduced to pools of resources to which the elected officials resort to in order to confront territorial competition. Furthermore, these studies perpetuate a very French tendency to dilute the urban issue and its potential specificities in the generic and unsatisfying category of the 'local'. This lack of differentiation between local spaces is, once again, indicative of the difficulty of the social sciences in France when studying local policies to recognise the spaces and urban societies as realities that pose specific problems to the political powers.

This difficulty that the social sciences, and more specifically French political science, have in conceiving cities as potentially specific political spaces contrasts with the precocious constitution of the city, of the urban policy/policies as specific objects in the North American social sciences field. In the United States, the different forms of the state's construction, the rapid establishment of towns as 'separate' political spaces (given the anti-urban ideology of the WASP majority, the greater presence of ethnic minorities, the establishment of urban political machines, etc.) have led researchers not only to pay more attention to the specificities of political life in the city, but also to the specific means of engaging in and implementing urban policies within a social context marked by racial cleavages, the role of large companies and the logic of property or, furthermore, the exodus of the white middle classes to the suburbs and the secessionism of privileged enclaves. Even if this way of postulating the specificity of urban policy/policies may have led to a certain isolation of the urban politics community within the field of North American political science (Sapotichne et al. 2007), it has, nevertheless, had the merit of repositioning urban societies and their specific characteristics at the centre of studies on urban politics.

Governance and Urban Regimes

The two decades since 1990 witnessed an evolution in the way French cities were governed and the urban policies made on the one hand, and an evolution in the way social sciences looked at these subjects on the other. The most striking sign of this change in outlook is the dissemination of an approach in terms of urban governance.

Today, evolutions in the forms of governance of French towns justify the taking of a new theoretical step that looks at cities as spaces governed by regimes.

From SPAL to Urban Governance

The *système politico-administratif local* has had its day! The way the social sciences view local political spaces must, therefore, change. While a certain number of significant traits of this 'system' remain – in particular the fragmentation of the *communes* and the dominant role played by local and regional government executives, many other aspects of these local political spaces have changed considerably.

The first of these changes concerns the widening of local government competences and the reinforcement, in particular regarding large cities, of their technical and political ability to conceive and implement public policies. The decentralisation legislation certainly enacted a distribution of competences between levels of territorial communities but they also left unchanged a principle of 'general competence' which has been in effect since the nineteenth century and which, in effect, provides the scope for intervention in a great many fields across every possible level. Large towns have, therefore, been able to develop their intervention in areas such as economic development, social assistance, housing, culture, but also in the building and maintenance of university and judicial structures. During the years following decentralisation, these innovations were often the work of enterprising mayors, or indeed adventurers, sometimes acting in legally grey areas and, in some cases, have even subsequently faced court cases. Later, certain public sectors underwent restructuring thanks to the advent of better trained urban professional bureaucracies, thanks also to the constitution of professional networks allowing for, on a national level, the circulation of expertise on urban policies, and thanks equally to the refinement of internal and external financial control mechanisms. This reinforcement of the capacity for action on the part of communities, and in particular of larger cities, means that today local and regional governments are responsible for more than 70 per cent of public investment expenditure.

The second major factor of change is the increase in power of the inter-municipal cooperation institutions, in particular in the urban context. Even before the decentralisation laws, because of the high level of fragmentation of the municipality map, some of them regrouped within these institutions, enabling them to jointly manage a number of services, networks and utilities (transport, waste collection and treatment, water supply and sanitation) and to perform a certain number of strategic functions (urban planning, economic development, housing, etc.) on a larger scale, all without jeopardising the legitimacy of the communal space as the basic unit of local democracy (Baraize and Négrier 2001, Desage 2005). Therefore, France has followed the general tendency to favour the metropolitan governance formulas of what has been termed 'neo-regionalism' (Savitch and Vogel 2009) consisting of a middle way between the solution of generalised competition between communities advocated by the *public choice* supporters and metropolitan government formulas which consist of transferring essential functions and resources from the communes to supra-communal institutions governed by directly elected councillors. The political functioning of those new institutions is of an inter-governmental type and does, in fact, attribute a prominent role to the mayors of each municipality in inter-communal policy negotiations. The state has played an essential role in the acceleration of this inter-communal regrouping process playing up financial incentives as well as threats of sanction. However, one must not neglect the role that shared and established working habits, acquired previously within a framework of more malleable formulas of inter-communal cooperation, play in this process as well as the socialising effect of inter-communal cooperation devices for planning and visioning, which have led elected entities to incorporate and internalise a need for cooperation.

The third element of change impacting on town government is the progressive delegation by the state of the 'strategic direction' guiding the production of urban policies towards local actors. At the beginning of the 1980s, the decentralisation laws gave a legal frame to the retreat of the 'dirigiste', planning and controlling central state and to the inexorable rise of the largest cities as political actors. The end of the de Gaulle era and the beginnings of the 1974 crisis marked the end of a development cycle for the state as planner. DATAR lost its influence and the central state as a whole experienced increasing difficulties in formulating a new project for the organisation of the national territory (Béhar and Estèbe 1999). The revival of planning in a regionalised and contractual format continued after decentralisation had little impact. Henceforth, it was to be cities and regions charged with the task of elaborating development strategies. Territorial projects can therefore only be local. That is how, throughout the last three decades, we have seen towns and inter-communal institutions expand their services to encompass economic development and carry out economic policies (Le Galès 1993), revive spatial and strategic planning (Padioleau and Demeestere 1991), develop international relations and marketing policies and submit applications to organise major events. Therefore, French cities and urban societies have acquired the capacity to think independently and project themselves in an environment that has become more uncertain and more competitive, in short they have learned to elaborate visions and projects (Pinson 2009).

Finally, the last factor of evolution concerns the process of pluralisation that has affected French urban policy networks, in other words, the multiplication of actors intervening in the fabric of urban policies and city governance, plus the consequent dispersion of resources for action (Pinson 2006). Whereas these action systems had remained relatively simple up until the 1970s and were essentially concentrated around state field services, elected officials and, in a secondary way, municipal services, they later became much more complex, with the increase in power of semi-public agencies and companies that gravitated around the administrations, taking over new fields of action, with the growing power of local branches of major real estate groups and urban utilities, with the new energy of certain organisations that represent economic actors or, furthermore, with the progressive autonomy of the state's local administrations, such as ports or universities, that play an increasingly important role in local projects and local development strategies. This proliferation of actors was accompanied by a dispersion of resources – financial, political, of expertise – permitting these actors to influence the drafting and the implementation of urban policies. To make that pluralisation compatible with the constitution of a capacity for collective action, urban political actors had to innovate in terms of public action instruments and we saw in the 1990s and 2000s the use of forms of action based on negotiation, deliberation and repetition, such as contracts, projects and other prospective initiatives (Gaudin 1999, Pinson 2004) becoming generalised.

In the French academic field, a new generation of researchers tried, in the 1990s, in line with the already quoted works by Lorrain, and often inspired by North American studies, to make the analysis of local political spaces more sensitive to the specificity of urban societies, to the logic of pluralisation gripping them, but also to the possible effects of the transformations of the capitalism. Lorrain (1990, 2002) documented the growing power of 'infrastructure firms' such as Générale des Eaux (that became Veolia) and Lyonnaise des Eaux (that became Suez). In his studies of local policies for economic development, Le Galès (1993) focused on the importance of the 'locality effect' and more precisely on the role of local mobilisations and interactions between urban social groups to explain the content of these policies. Afterwards, inspired by the American political economy approach to city governance, he proposed the renewal of the approach of urban policies using the term 'governance', a notion able to encapsulate the phenomena of the pluralisation of urban actors in conjunction with the decentring of public actors and regulations in urban production and management and instrumental innovation in urban policies (Le Galès 1995). Borraz has explored, with an organisations sociology approach, the phenomenon of organisational proliferation in urban action systems and the renewal of practices by elected urban officials to cope with this new reality (Borraz 1998, Borraz and John 2004). More recently, researchers have tried to apply neo-Marxist and regulationist approaches to French cities, in particular by trying to test the hypothesis of a neo-liberalisation of the forms of urban governance (Béal and Rousseau 2008, Jouve 2009).

The majority of these studies share a certain number of findings and postulates that lead them to agree on the fact that French towns are no longer governed as they were in the 1960s and that a series of evolutions require the renewal of analytical frameworks. Firstly, despite the traditional weight of the central state in economic activities, the transformations of capitalism and of productive systems have had a strong impact on the relationship between the economy and the space, in particular in the sense of a localisation of economic interactions. These transformations, documented by geography and economic sociology, as well by the Ecole de la Régulation (Benko and Lipietz, 2000, Boyer 1986, Veltz 1996), have re-evaluated the role of local spaces as purveyors of positive external factors, such as reduced transactional costs and inducing the development of interactions between economic actors and urban government organisations. Whereas their role in economic development had become marginal within capital concentration processes, the constitution of large national groups under the guidance of the state and its interventionism (Levy 1999), the local communities and in particular the large towns and regions, have become key actors in economic development. This, inevitably, has an impact on the practices and on the socialisation modes of the elected entities and the bureaucrats and, therefore, on the alliances and power relationships at the heart of towns. I have documented, based on British and Italian cases, the effects of the intensification of the connections with and between actors, who are purveyors of urban public action resources (economic actors, social groups, themselves purveyors of expertise), on the practices and forms of sociability of elected urban entities and on their relationships with the electoral clientele and on the work involved in maintaining that clientele (Mattina 2007, Pinson 2007). To summarise, the daily immersion in policymaker networks and the mobilisation of support in the heart of social groups that hold the resources for action tends to replace the structuring of social groups whose only resource is the vote.

The second finding shared by these studies is the increasing complexity of French urban societies. The studies of organisations sociologists or of the political scientists interested in local realities pictured urban societies of the 1960 and 1970s as being relatively simple, organised around several social groups that were relatively easy to circumscribe. This simplicity was mirrored in the organisation of municipal administrations which were still fairly meagre up until the 1970s. The simplicity of town social structures had the advantage of facilitating the work of mobilising electoral support. All the elected entity had to do was to obtain the support of organisations (the church, trade unions, associations, etc.) and of opinion leaders judged able to 'hold' certain segments of local society and tap into their electoral support. The simplicity of the administrative structure facilitated the conception of urban policies which, in essence, were in any case the responsibility of state services. Borraz (1998, 2000), Hoffmann-Martinot (1999) or Jouve and Gagnon (2006) demonstrated in their studies how this situation evolved. The increasing weight of the salaried middle classes, of immigration and the amplification of mobility and internal migrations complexified the social structures of French cities, making more uncertain any possibility of structuring urban societies into effective electoral clienteles by means of well identified relays. The organisational proliferation which marked the structuring of urban and metropolitan bureaucratic structures in the 1980s and 1990s reinforced interdependence between organisations, between the state and the different levels of local communities, considerably complicating the production of public policies.

Verifying this increasing complexity leads directly to a third finding: that of the transformation in the means of producing urban policy. The multiplication of policy issues and fields of intervention for urban administrations, the proliferation of actors and organisations, the dispersion of resources for initiatives made action devices more complex to manage. The importance acquired by urban policies and their increasing complexity led to the instrumental innovations mentioned above, but they also modified the balance of power at the heart of urban societies. Municipal and in particular inter-communal techno-structures have clearly changed, gaining influence at the expense of state representatives. Municipal and inter-communal executives, as well as their entourage (office, experts) have progressively marginalised councils. Social groups holding strategic resources for urban policies have increased their capacity to access decision-making processes and influence the latter to the detriment of groups whose only resource is the vote. Equally, the increasing complexity of the production of urban politics has deeply transformed the role of the top elected officials, in particular that of 'mayors' whose role is ever less structured by activities encompassing mediation, access to the centre to obtain the adjustment of rules or the maintenance of electoral clienteles, and increasingly by activities related to the production of visions and projects that lend sense and direction to policy networks and to the mobilisation of scarce resources for policies. To qualify this transformation of the forms of the urban political profession and externalise the difference in particular regarding the figure of the '*notable*' entity, the term 'leader' has been judged appropriate by certain authors (Borraz and John 2004, Le Galès 2003, Pinson 2007).

If the use they make of the notion of 'governance' can vary, however, the scholars quoted thus far can be all located in the 'governance research field' (Pinson 2003). They all come from a research perspective that seeks to trace the ways in which the governance of towns has changed. Receptive to studies on urban economic policies, they are observant of the potential effects of transformations in capitalism on the logic of urban policy production and on the governance of towns. They are also concerned with the repercussions that these changes may have on the sphere of policies and politics, on electoral competition and on the modalities of political socialisation of urban social groups. Today, these authors and their studies have managed to structure French academic debate regarding local policies around the notion of 'governance', which allows the incorporation of a certain number of changes that have occurred both in the economic policies of towns, in urban public action and in the forms political activities take in towns. Within the framework of that debate, they often confront researchers more sensitive to factors of stability and that frequently limit their approach to politics, to political competition types, to the construction of eligibility and electoral strongholds, and who contest the heuristic added value of that notion of 'governance'.

From Centre–Periphery Relations to Analysis of Urban Regimes

It is now time to take a new step forward in the approach to studying the forms of governance of French towns. A good way of doing that would be to look at these towns with an approach adopted by North American studies on urban regimes. The main merit of these works is to shift the way one looks at towns and no longer consider the vertical relations between urban actors on the one hand, and local and central state representatives on the other, as evidently structuring the forms that urban governance actually take. On the contrary, the regimes approach postulates that horizontal relations, in other words the conflicts, cooperation and alliances that link actors, groups and organisations physically present in the town, can be seen as factors that explain, as much or even more, the actual form of urban governance.

The main tendency in French studies on local government up until the 1990s has been to minimise the role of these horizontal relations. And quite rightly in most cases! During the period between 1950 and 1970, the central state was the main actor in urban policies and the major role of elected entities was to attract the attention of the state to the problematic specificities of their town or to negotiate the way national programmes would be locally implemented. The specific relationships between local actors and social groups seem to have had no impact on the content of urban policies and the way they were implemented. In political sociology studies of localities, those horizontal social networks are almost invisible: the '*notable*' collects electoral supports in a relatively simple fashion. These supports consent access to the national centre, the sole place where real political careers are made and where the resources to carry out urban policies can successfully be obtained. In both cases, what happens locally only makes sense in relation to the centre. In both cases, local spaces feature as dominated political spaces. In both cases, these subordinate spaces are drowned in a generic category, the 'local space', which dilutes the socio-economic specificities of the places and does not allow for differentiating the dynamics characteristic of towns.

The increasing complexity of urban societies, the pluralisation of political systems, the fact that an increasing number of actors participate in designing and implementing urban policies and that, at the same time, the state provides fewer resources for implementing those policies successfully, render this centre–periphery approach less and less feasible. In contrast, the governance approach allows for the integration of the economic context and the evolution of urban policymaking in the analysis of urban politics. As for the urban regime approach, it enables paying even greater attention to the horizontal interactions between urban actors of different natures and to urban coalition making processes.

Developed in the United States by Clarence Stone (1989, 1993), transposed to the United Kingdom by Harding (1997, 2000), the urban regimes approach has, in France, rarely been taken up, with the notable exception of the studies by Dormois (2006). This approach consists of a synthesis of urban political economy works on the one hand and of the intuitions of pluralism on the other. The first, mainly represented by the works of Logan and Molotch on 'growth machines' (1987), insists on the fact that the political life of North American cities is largely structured around conflicts opposing local social groups to the urgency, the pace and forms of economic development in towns. With the retraction of federal programmes, the growth issue has become central in the politics of American cities and triggers most political conflicts. In these conflicts, more often than not, coalitions or growth machines are constituted and end up dominating the political agenda. These 'growth machines' are constituted, on the one hand, by the elected urban officials who need the financial support of the local business elites in order to be re-elected and, on the other, by companies, real estate promoters, important land owners, local media, etc. who have a vested interest in the town's growth and try, from the beginning, to have a say regarding urban policies, and in particular on urban planning. The application of the 'growth coalition' model to French cities and, more generally, to European towns poses a problem because the large companies and what Logan and Molotch call 'structural speculators' (promoters and large private landowners) do not have the weight they have in American towns, given the greater fragmentation of soil ownership and the weight of public or semi-public land ownership (municipality, public enterprises, hospitals, religious orders, etc.) (Harding 1994). On the other hand, the still significant weight of publicly redistributed incomes on local income makes European elected entities much less dependent on the great local private interests (Davezies 2008). However, the 'growth machines' approach has the advantage of looking more attentively at the major developments that have occurred in the governance of French towns. First of all, be they on the right or on the left politically, the majority of urban governments put urban growth as their major priority. Naturally, this growth is conceived as having to be 'reasonable' or 'sustainable' and cannot compromise the essential amenity that constitutes the town's standard of living framework, but is often presented as an insurmountable horizon. Subsequently, one observes in the coalitions mobilised around urban politics in France, as in the American 'growth machines', the involvement of institutions and local agencies such as the local media, the universities and also cultural institutions who all have a vested interest in growth. The pluralist approaches (Dahl 1971), whose studies on the regimes recognise heredity, in turn insist on the great number of actors and organisations intervening in urban governance and the dispersion of resources (financial, expertise and, in particular, political legitimacy). These approaches insist more heavily than political economy approaches do, on the considerable importance that actors and political institutions can have, even in a capitalist context that grants a systemic advantage to those who hold capital. That weight can be expressed by the capacity to mobilise electoral support for a development project or by the grip officials can have on the regulatory framework through which urban development has to go.

The synthesis proposed by the urban regimes approach takes into consideration, firstly, that in a capitalist economy there is always a strong possibility that there will be a 'systemic bias' in urban politics favouring economic interests. That has always been the case in the United States, where towns collect their main fiscal revenues from the presence of companies. It is also increasingly the case in French towns where, although the resources distributed by the state or from social redistribution systems play a greater part in the local economy, it is nevertheless true that the elected officials have their eyes fixed on the departures and the arrivals of companies. These pressures exerted on urban government in a capitalist regime lead to, according to these authors, the constitution of government coalitions, 'urban regimes', bonding government institutions and certain major economic interests in a lasting relationship. Furthermore, the regimes theory is more interested in the nature of the relationships linking the actors involved and draws conclusions close to those of the pluralists. The systems of urban actors of American towns are characterised by a high level of complexity, fragmentation and interdependence among actors and by the dispersion of resources. The central challenge of urban governance – it is no longer time for the elected entities to decide unilaterally, but to 'introduce enough cooperation between the various elements that constitute the community in order to actually make things happen' (Stone 1989: 227). Thus, the situations that pluralism entails render futile the resorting to hierarchical leadership as a coordination mode. Cooperation comes instead from the capacity of elected officials and their administration to interest actors in their urban policies. This capacity comes from the stabilisation of governance arrangements, from the relations of trust and reciprocity between autonomous and interdependent actors. The relations are self-perpetuating thanks to modes of action and interaction that do not question the autonomy of actors. Little by little, the stabilisation of these trusting relationships generates a capacity to cooperate and act collectively. The regime becomes the matrix for the city's capacity for collective action. Identified as such by coalition members, the regime becomes an asset worthy of being protected by each participant.

Finally, the picture drawn by the theory of urban regimes seems to be able to describe, with some subtle and occasional differences, the situation of French cities and their modes of governance. These towns are subject, more than previously, to the transformations of capitalism and are less protected by the state's umbrella and by Keynesian policies. The challenges of economic development, competitiveness and attractiveness have become central issues. Urban elites are required to define a position in territorial competition whereas the state provides almost no expertise or important guidelines on the subject. Therefore, projects are defined locally. On the other hand, the growing dependence with regard to private investment makes the urban elites ever more attentive to economic actors, which means that, even when they are not physically present in French urban regimes, their supposed interests are broadly taken into consideration. As urban societies are increasingly required to build up their own projects, they simultaneously become more polycentric and pluralist. As the actors representing the state began to take a back seat, urban actors systems became occupied by an increasing number of local groups and organisations. As the links with the centre lose their strategic character, the most evident challenge of urban governance becomes that of making the pluralism of urban actor systems prosper, of making it compatible with the constitution of a capacity for action. The challenges involved in mobilisation, coordination, the building of stable coalitions and the constitution of the town as a collective actor have become central issues. Innovation in terms of tools for urban public actions (associations, contracts, projects, pacts, charters), the emphasis on the major mobilising initiatives (large urban projects, strategic planning, major events, etc.) (Pinson 2002) are all relevant elements revealing the importance taken on by the densification and management of horizontal relations in urban governance.

Conclusion: Southern European Urban Regimes?

The regimes approach has considerable merits. It allows for the reconciliation of political economy approaches and policy analysis approaches more sensitive to the inherent complexity of drafting and implementing urban policies and to the interdependent dynamics between actors. This approach also incorporates features that French political science very often neglects: the fact that cities are socially, economically and politically complex phenomena, polycentric ecosystems in which questions regarding the coordination and mobilisation of actors around public policies and the integration of social systems are always emerging differently from that able to be observed somewhere else. Finally, because this approach perceives how coalitions within urban governance will always be dependent on the locally prevailing 'social equation', the urban regimes approach is sensitive to the question of differentiation between the forms of urban governance within the context of the same national governance. Furthermore, what transnational studies on urban governance today demonstrate (Le Galès 1993, Pinson 2009, Sellers 2002) is that the infra-national variations can be more important than international variations.

No doubt one must henceforth try to qualify what could be the unique characteristics of European urban regimes, in particular in Southern European. Southern European regimes grant a much more important place to actors and organisation which Stone tends to neglect: technical networks, planning professionals, agencies, public and semi-public enterprises and their staff, infrastructural managers or cultural institutions. These actors often occupy a more decisive place than economic actors. Furthermore, public actors and institutions or semi-public institutions have, undoubtedly, a larger role given the historic importance that political regulations and welfare systems have. Top urban elected officials undoubtedly occupy a choice place in these regimes. They play an essential role in the mobilisation of the urban society components and in the construction of mobilising images and messages. One might perhaps propose that the reflex and interest in organising and mobilising oneself at the urban level is, without doubt, less rooted in groups that compose European urban societies than in their North American equivalents. In Europe, the state, even when withdrawn, is, above all, a 'state of mind'.

References

Abélès, M. 1989. *Jours Tranquilles en 89: Ethnologie Politique d'un Département Français*. Paris: Odile Jacob.

Baraize, F. and Négrier, E. eds. 2001. *L'Invention Politique de l'Agglomération*. Paris: l'Harmattan.

Béal, V. and Rousseau, M. 2008. Néolibéraliser la ville fordiste. Politiques urbaines post-keynésiennes et re-développement économique au Royaume-Uni: une approche comparative. *Métropoles*, 4, 160–202.

Béhar, D. and Estèbe, P. 1999. L'Etat peut-il avoir un projet pour le territoire? *Les Annales de la Recherche Urbaine*, 82, 80–91.

Benko, G. and Lipietz, A. eds. 2000. *La Richesse des Régions*. Paris: PUF.

Borraz, O. 1998. *Gouverner une Ville: Besançon, 1959–1989*. Rennes: Presses Universitaires de Rennes.

Borraz, O. 2000. Le gouvernement municipal en France. Un modèle d'intégration en recomposition. *Pôle Sud*, 13(1), 11–26.

Borraz, O. and John, P. 2004. The transformation of urban political leadership in Western Europe. *International Journal of Urban and Regional Research*, 28(1), 107–20.

Boyer, R. 1986. *La Théorie de la Régulation: une Analyse Critique*. Paris: La Découverte.

Brenner, N. 2004. *New State Spaces: Urban Governance and the Rescaling of Statehood*. Oxford: Oxford University Press.

Briquet, J. and Sawicki, F. 1989. L'analyse localisée du politique. *Politix*, 2(7), 6–16.

Cadiou, S. 2009. *Le Pouvoir Local en France*. Grenoble: PUG.

Castells, M. and Godard, F. 1974. *Monopolville*. Paris/La Haye: Mouton.

Chédeville, A., Le Goff, J. and Rossiaud, J. 1980. *Histoire de la France Urbaine*. Paris: Seuil.

Dahl, R. 1971. *Qui gouverne?* Paris: A. Colin.

Davezies, L. 2008. *La République et ses Territoires: La Circulation Invisible des Richesses*. Paris: Seuil.

Desage, F. 2005. *Le Consensus Communautaire contre l'Intégration Intercommunale. Séquences et Dynamiques d'Institutionnalisation de la Communauté Urbaine de Lille (1964–2003)*. Thèse pour le doctorat en science politique, Université de Lille 2.

Dormois, R. 2006. Structurer une capacité politique à l'échelle urbaine. Les dynamiques de planification à Nantes et à Rennes (1977–2001). *Revue Française de Science Politique*, 56(5), 837–67.

Elias, N. 1975. *La Dynamique de l'Occident*. Paris: Calmann-Lévy.

Fontaine, J. and Le Bart, C. eds. 1994. *Le Métier d'Élu Local*. Paris: l'Harmattan.

François, B. 2008. *Le Régime Politique de la Vè République*. Paris: La Découverte.

Garraud, P. 1989. *Profession Homme Politique: la Carrière Politique des Maires Urbains*. Paris: l'Harmattan.

Gaudin, J.P. 1985. *L'Avenir en Plan: Technique et Politique dans la Prévision Urbaine, 1900–1930*. Seyssel: Champ Vallon.

Gaudin, J.P. 1999. *Gouverner par Contrat: l'Action Publique en Question*. Paris: Presses de Science Po.

Grémion, P. 1976. *Le Pouvoir Périphérique. Bureaucrates et Notables dans le Système Politique Français*. Paris: Seuil.

Harding, A. 1994. Urban regimes and growth machines: toward a cross-national research agenda. *Urban Affairs Quarterly*, 29, 356–66.

Harding, A. 1997. Urban regimes in an Europe of the cities? *European Urban and Regional Studies*, 4(4), 291–314.

Harding, A. 2000. Regime formation in Manchester and Edinburgh, in *The New Politics of British Local Governance*, edited by G. Stoker. London: Palgrave Macmillan, 54–71.

Hoffmann-Martinot, V. 1999. Les grandes villes françaises: une démocratie en souffrance, in *Democraties urbaines: L'état de la democratie dans les grandes villes de 12 pays industrialisés*, edited by O. Gabriel and V. Hoffmann-Martinot. Paris: l'Harmattan, 30–45.

Jobert, B. and Muller, P. 1987. *L'État en Action: Politiques Publiques et Corporatismes*. Paris: Presses Universitaires de France.

Jouve, B. 2009. Ville: Le grand retour de la pensée critique. *Place Publique. Nantes/ Saint-Nazaire*, 15 [available at: www.revue-placepublique.fr/Sommaires/ Articles/retourcritique.html].

Jouve, B. and Gagnon, A. eds. 2006. *Les Métropoles au Défi de la Diversité Culturelle*. Grenoble: Presses Universitaires de Grenoble.

Lagroye, J. 1973. *Société et Politique. Jacques Chaban-Delmas à Bordeaux*. Paris: Pedone.

Le Galès, P. 1993. *Politique Urbaine et Développement Local: une Comparaison Franco-Britannique*. Paris: Editions l'Harmattan.

Le Galès, P. 1995. Du gouvernement des villes à la gouvernance urbaine. *Revue Française de Science Politique*, 45(1), 218–28.

Le Galès, P. 2003. *Le Retour des Villes Européennes: Sociétés Urbaines, Mondialisation, Gouvernement et Gouvernance*. Paris: Presses de Sciences Po.

Le Lidec, P. 2001. *Les Maires dans la République. L'Association des Maires de France, Élément Constitutif des Régimes Politiques Français depuis 1907*. Thèse de doctorat de science politique. Université de Paris I.

Levy, J. 1999. *Tocqueville's Revenge: State, Society, and Economy in Contemporary France*. Cambridge, MA: Harvard University Press.

Logan, J. and Molotch, H. 1987. *Urban Fortunes. The Political Economy of Place*. Berkeley: University of California Press.

Lorrain, D. 1989. La montée en puissance des villes. *Economie et Humanisme*, 305(6), 6–21.

Lorrain, D. 1990. Le modèle français de services urbains. *Economie et Humanisme*, 312, 39–58.

Lorrain, D. 1991. De l'administration républicaine au gouvernement urbain. *Sociologie du Travail*, 33(4), 461–84.

Lorrain, D. 2002. Capitalismes urbains: la montée des firmes d'infrastructures. *Entreprises et Histoire*, 30, 7–31.

Mabileau, A. 1991. *Le Système Local en France*. Paris: Montchrestien.

Martin, R. 1989. The new economics and politics of regional restructuring: the British experience, in *Regional Policy at the Crossroads: European Perspective*, edited by L. Albrechts. London: Jessica Kingsley, 27–51.

Mattina, C. 2007. Changes in clientelism and urban government: a comparative case study of Naples and Marseilles. *International Journal of Urban and Regional Research*, 31(1), 73–90.

Mény, Y. 1987. France, in *Central and Local Government Relations*, edited by M. Goldsmith and E. Page. London: Sage, 88–106.

Mény, Y. 1992. La république des fiefs. *Pouvoirs*, 60, 17–24.

Orr, M. and Johnson, V. 2008. *Power in the City: Clarence Stone and the Politics of Inequality*. Lawrence: University Press of Kansas.

Padioleau, J. and Demeestere, R. 1991. Les démarches stratégiques de planification des villes. *Les Annales de la Recherche Urbaine*, 51, 28–39.

Petaux, J. 1982. *Le Changement Politique dans le Gouvernement Local: le Pouvoir Municipal à Nantes*. Paris: Pedone.

Pinson, G. 2002. Political government and governance: strategic planning and the reshaping of political capacity in Turin. *International Journal of Urban and Regional Research*, 26(3), 477–93.

Pinson, G. 2003. Le chantier de recherche de la gouvernance urbaine et la question de la production des savoirs dans et pour l'action. *Lien Social et Politiques*, 50, 39–55.

Pinson, G. 2004. Le projet urbain comme instrument d'action publique, in *Gouverner par les instruments*, edited by P. Lascoumes and P. Le Galès. Paris: Presses de Science Po, 199–233.

Pinson, G. 2006. Projets de ville et gouvernance urbaine. Pluralisation des espaces politiques et recomposition d'une capacite d'action collective dans les villes europeennes. *Revue Française de Science Politique*, 56(4), 619–51.

Pinson, G. 2007. Gouverner une grande ville européenne: les registres d'action et de légitimation des élus à Venise et Manchester. *Sciences de la Société*, 71, 89–114.

Pinson, G. 2009. *Gouverner la Ville par Projet: Urbanisme et Gouvernance des Villes Européennes*. Paris: Presses de Sciences Po.

Politix. 2001. Le temps des mairies. *Politix*, 53.

Rondin, J. 1985. *Le Sacre des Notables. La France en Décentralisation*. Paris: Fayard.

Sellers, J. 2002. *Governing from Below: Urban Regions and the Global Economy*. Cambridge: Cambridge University Press.

Sapotichne, J., Jones, B. and Wolfe, M. 2007. Is urban politics a black hole? Analyzing the boundary between political science and urban politics. *Urban Affairs Review*, 43(1), 76–106.

Savitch, H. and Vogel, R. 2009. Regionalism and urban politics, in *Theories of Urban Politics*, edited by J. Davies and D. Imbroscio. Los Angeles: Sage, 106–24.

Stone, C. 1989. *Regime Politics: Governing Atlanta, 1946–1988*. Lawrence: University Press of Kansas.

Stone, C. 1993. Urban regimes and the capacity to govern: a political economy approach. *Journal of Urban Affairs*, 15(1), 1–28.

Tilly, C. and Blockmans, W. eds. 1994. *Cities and the Rise of States in Europe, A.D. 1000 to 1800*. Boulder: Westview Press.

Veltz, P. 1996. *Mondialisation, Villes et Territoires. L'Économie d'Archipel*. Paris: PUF.

Weber, M. 1982. *La Ville*. Paris: Aubier.

Worms, J.P. 1966. Le préfet et ses notables. *Sociologie du Travail*, 3(66), 249–75.

Chapter 8

Integrated Urban Interventions in Greece: Local Relational Realities Unsettled[1]

Ioannis Chorianopoulos[2]

This chapter portrays the urbanisation and governance profile of Greek cities and explores their capacity to respond to EU calls for enhanced competitiveness and synergistic policy-making. The distinctiveness of the Greek urban example, it is argued, is not well adjusted to European spatial priorities and neo-corporatist modes of intervention. In this light, key territorial restructuring reforms were launched in the 1990s, bearing the influence of EU urban policy discourse. The impact of these changes on local realities is explored through case-study research, focusing on cities that participated in the two phases of the URBAN Community Initiative: Keratsini-Drapetsona (1994–9) and Iraklion (2000–6). The chapter concludes by discussing the prospects and challenges opening up for local development as a result of the latest state spatial restructuring reform in Greece (2010), the directions of which alter radically the geographical dimensions of state power.

Introduction

The competitive position of the European economy against those of the United States and Japan deteriorated sharply during the late 1980s and early 1990s.[3] This, together with lower trade barriers through reductions in the rate of customs duties, already agreed at that time in the Uruguay Round of the GATT multilateral trade negotiations, prioritised the issues of European economic adaptability and competitiveness in the Maastricht Treaty (CEC 1994a: 3). The White Paper on 'Growth, Competitiveness and Employment' highlighted the importance of flexible and decentralised policies as a means to enhance European economic potential.

---

1 An early version of this text was published in 2010 in *Análise Social. Revista do Instituto de Ciências Sociais da Universidade de Lisboa*, 197, 739–56.

2 University of the Aegean; Department of Geography. University Hill; Mytilene 81100 (Greece). i.chorianopoulos@geo.aegean.gr.

3 As illustrated by shares in export markets in general, but also from research and technical development and innovation exports and development of new products (CEC 1993a: 9).

This policy perspective centres on the role of the local political level, identified as the most appropriate in generating cooperation processes and endogenous, place-specific, development paths[4] (CEC 1993a: 9).

The increased responsibilities of the local level illustrate the gradual adaptation of Community policies to the rapidly changing economic circumstances apparent since the 1980s. The upgraded role of the local level in pursuit of national and European economic objectives is approached in Community documents through the analytical framework of 'globalisation and the shift from industries to services [that] … enhanced the importance of space for economic development, [and] … reinforced the potential of cities as autonomous creators of prosperity' (CEC 1997: 6, 8). Urban areas, in this context, are portrayed as 'the main source of prosperity [as] … they contribute disproportionately more to regional or national GDP compared to their population' (CEC 1997: 4). In terms of social disparities, the enhanced importance of cities in Community policies is approached on the basis of concentrated social problems in urban areas linked to low educational attainment rates, high (above European averages) urban unemployment rates, and social exclusion phenomena. Such concerns have kept a certain momentum ever since and currently mark EU spatial policy prioritisations. Urban policy is now an explicit part of the National Strategic Reference Frameworks (CEC 2007), a development that underscores the increased importance of the EU stance to the prospects of European cities. Within this framework, this chapter focuses on Greece.

Greece joined the EU in 1981. The extent to which the country was not in a position to participate in an enhanced competitive context is indicated by the underdeveloped state of key aspects of the economic environment, such as physical infrastructure provision and workforce qualifications.[5] The political prioritisation of the acceleration of economic integration aggravated concerns about the capacity of the country to adjust along competitiveness lines. Greek cities were gradually called on to participate in this drive for competitiveness. The results, however, have not been encouraging. With the exception of Athens, Greek cities lag behind in critical competitiveness indicators (DATAR 2003), exhibit poor economic performance and fall short of participating in the European urban core, the area that encompasses the majority of economically dynamic European cities (ESPON 2006). On the contrary, it appears that the enhanced inter-urban competition that followed the Single European Act (1986), and the subsequent diminution of protectionist barriers to trade and investment, exposed the traits of

4 As indicated by the White Paper 'it is the local level at which all the ingredients of political action blend together more successfully' (CEC 1993a: 9).

5 The extent of the motorway network, for instance, was less than 10 per cent of the Community average. In terms of workforce qualification levels, the rate of adults who had not completed education beyond primary school level was one in every two – as opposed to the one in five average across the Community, while the country had only 20–25 per cent of the Community's average of persons employed in Research and Technology Development (RTD) (CEC 1994b: 10–11).

an urban system which faces embedded difficulties in grasping the development opportunities opened up by EU integration. EU urban policy aims specifically to remove obstacles to growth, strengthen competitiveness and promote sustainable development (CEC 2006a). This chapter explores the responses of Greek cities to EU urban policy and it is organised in three parts.

The first part discusses the ways in which cities were defined as a policy problem by the EU, also highlighting the traits of EU urban policy. The second section focuses on Greece. It draws attention to the particularity of the economic and socio-political realities of the post-war period of rapid urbanisation, underscoring the distinctive context in which Greek cities were called on to promote restructuring strategies. This is followed by a brief description of recent politico-administrative attempts seeking to enhance the regulatory attributes of the local level. The third part presents the governance responses of Greek cities to competitiveness-oriented prioritisations. The case-study cities are Keratsini-Drapetsona and Iraklion, empirical analysis spanning more than a decade of urban interventions. Research findings suggest significant transformations in the way that urban issues are approached, defined and administered in the country – changes that show the influence of the EU urban policy discourse. The distinctiveness of the urbanisation and governance characteristics of Greek cities is, however, still present, shadowing EU policy prioritisations and questioning the relevance of the EU urban approach to the local context. The chapter concludes by presenting current urban policy directions in Greece, as defined by the latest state spatial restructuring reform (2010), and the incorporation of the urban dimension into the National Strategic Reference Framework (2007–13).

Europe and the City

Following a period of brief experimentation with a series of innovative urban intervention schemes, the increased importance attached by the EU on the urban level was reflected in the Maastricht Treaty (CEC 1992a). The amended framework regulating the responsibilities of the Community and national and local level authorities in structural policies re-affirmed the 'subsidiary' character of the Community's involvement at the national level. In this light, EU urban policy priorities and programmes were to be filtered at the national level, adapting them to local political-administrative structures. The subsidiarity concept, however, as adopted in Maastricht, also recognised the enhanced role of the local level in the pursuit of the Community's socio-economic objectives and – by promoting links between the local and Community levels – upgraded the role of local authorities within the EU spatial policy framework.[6] The rationale for the launch of decentralised policies also led to the extension of the 'partnership' principle to

6 The subsidiarity principle extends also to the sub-state level aiming to 'prevent sub-state centralisation' (Council of Europe 1994: 29). It is this interpretation of the subsidiarity

include 'competent authorities and bodies, [as well as] … economic and social partners, designated by the member State at national, regional, local or other level … in pursuit of a common goal' (CEC 1993b: 48).

The incorporation into a single EU programme of all the political authorities leading projects, as well as the relevant public and private sector interest groups, signifies the promotion of 'governance' structures as the basis of EU spatial policies. In this respect, collaboration aims at unlocking the local relational dynamic, triggering synergistic responses to local challenges. Such development-oriented neo-corporatist stances were noted in a number of European cities that managed to reverse de-industrialisation, exhibiting a restructuring record marked by high growth rates and population recentralisation (Camagni 2002, Cheshire 1995). The prioritisation of 'governance' policies as the selected policy form for the advancement of funding targets was applied explicitly at the urban level with the launch of the URBAN Community Initiative (1994–2006).

The introduction of URBAN was justified on the grounds of improved coordination of the structural instruments which, in accordance with the subsidiarity principle, would a) strengthen support for cities, especially those under Objectives 1 and 2; b) promote innovative policies and local level 'exchange of experience' networks; and c) extend the inclusion of the urban dimension in the formulation of various Community policies as a means to achieve more spatially focused economic development objectives (CEC 1994c: 1).

The key novel aspect of URBAN was the emphasis placed on the participation of local level interest groups in all programme phases (CEC 1998: 2). 'The inclusion of local authorities in partnership mechanisms [with] … economic and social bodies', was viewed as 'essential' for tackling urban deprivation and promoting economic competitiveness (Trojan 1998: 6). The relevance of the EU urban approach to Greek urban realities of administrative centralisation and limited experience of local collaborative arrangements is explored next. This analysis emphasises key qualities of the urban economic and socio-political space that distinguish the governance profile of Greek cities from the ideal-typical restructuring examples that inform EU urban policy.

Greek Urban Legacies

The debate over European urbanisation patterns correlates socio-economic transformation processes with distinct stages of urban spatial evolution (CEC 1992b: 71–2). The 1960–80 decline of population in urban agglomerations in Northern Europe, for instance, was approached as the spatial manifestation of industrial, employment and residential decentralisation processes (Berg et al. 1982: 29–36). Also, the emergence of a variation of urban development patterns

principle that is most relevant to the attempt to identify changing Community priorities in the field of spatial policies.

– with some urban regions displaying recentralisation tendencies since the 1980s – was correlated with the growth of service sector employment and the enhanced urban socio-economic embeddedness of economic activity (Cheshire 2006). In this framework, Greece represents a distinct example. Figure 8.1 displays the average urban growth rates for the period 1950–2005, also indicating the rate of urban population expansion as a percentage of the total population.

As noted in Figure 8.1, the wave of urbanisation experienced in Greece in the early post-war period was substantial and extensive. What is also of note in the above data is the continuous, albeit at lower rates, pattern of urban centralisation, a trend that raises the question of urban economies and 'pull' factors that originally drove and currently sustain the influx of population in cities. A look at the sectoral distribution of labour force during the last decades (Figure 8.2) sheds light on this trend.

At first glance, Figure 8.2 depicts the gradual 'modernisation' of the country's employment structures, from the dominance of the agricultural sector in the 1960s, to the key role of services in the 2000s (CEC 1992b: 67). Looking closer, however, the significant rates of service employment during the early period of urban centralisation are noteworthy (Adrikopoulou et al. 1992: 214, Williams 1984: 8). In fact, the working population engaged in service activities surpasses that of industry throughout this period. The moderate contribution of industry as a source of employment during the early period of urban growth indicates the constrained capacity of the sector to influence migration patterns. The narrow manifestation of internal economies of scale in industrial firms (Hudson and Lewis 1984: 200), as well as the few signs of external economies of localisation affecting the spatial pattern of industrial development,[7] further supports this argument. The limited role of industry as an initial 'pull' factor to urban centres and the precise nature of tertiary sector employment, dominated by public administration, tourist services and self-employment, points to the informality of migrant job expectations (Tsoukalas 1986: 184). Distinct spatial and socio-political landscapes are related to this particularity.

Urban Regulation

Despite the weak role of industry as an urban immigration 'pull' factor, post-war Greece displays a strong economic growth record,[8] also based on the gradual industrialisation of the economy (see Figure 8.2). The socio-institutional profile

7 By 1980, 57 per cent of the urban population was living in Athens (CEC 1992b: 62, 67). The concentration of urban growth in predominantly one urban centre is an indication of the absence of localisation economies – either based on firm specialisation or determined by the presence of raw material resources – influencing industrialisation (see Louri 1988: 434).

8 Average annual growth of Greek GNP per capita for the 1960–80 period was between 4 and 6 per cent, the highest amongst OECD members with the exception of Japan (Williams 1984: 8).

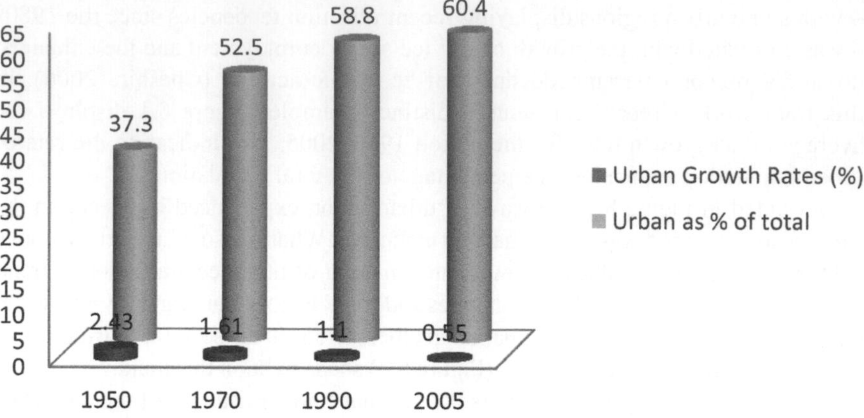

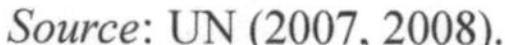

Figure 8.1 Urban population growth rates in Greece (1950–2005)

Source: UN (2007, 2008).

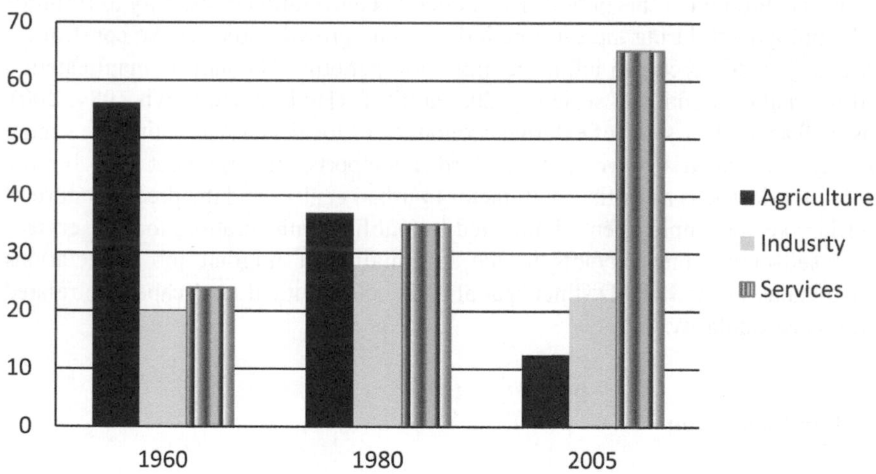

Figure 8.2 Sectoral distribution of labour force (1960–2005)

Sources: World Bank (1984: 259), OECD (2007: 30).

of the economy, however, differs substantially from the ideal-typical Northern European industrial development examples of the era. Post-war political realities in the country are time-framed by the Civil War (1945–9) and the military dictatorship (1967–74). During this time, trade-union representation was either restricted or utterly prohibited (1967–74), arresting the emergence of consensual wage relations and corporatist policies. Political legitimacy was claimed on the basis of urbanisation related economic expansion. It is in this context that national authorities tolerated and accepted informal economic activities, estimated at 25 per cent of GDP at factor cost throughout this period (Ioakimidis 1984: 42–3). Authoritarian rule and the paucity of national socio-political compromises were also reflected locally.

Political repression was combined in Greece with a mode of political-administrative centralisation aimed at arresting the surfacing of opposing political voices. Local authority income, for example, was collected on behalf of the local state by the Ministry of Finance, placing strict controls on local spending. More characteristically, the national authorities appointed mayors and public sector officials at the local level, impeding the emergence of local interests (Chlepas 1997). The limited regulatory experience of the local level in the post-war period influenced the direction of changes attempted with the re-establishment of democracy (1975), suggesting path-dependent endogenous development constraints.

The re-introduction of local elections (1975) was accompanied by an expansion of local level bureaucracies and functions. In the absence of an active local socio-political scene, however, the new national political scene exerted strong influence on local regulatory traits. Local authority financial competences, for instance, did not increase substantially and did not devolve. More than 80 per cent of local income is still levied centrally and is subsequently distributed at the local level on the basis of a number of rigid, population-related, criteria (Lalenis and Liogkas 2002: 443). Such inflexibility renders the financial activity of Greek local authorities last amongst the EU-15 (Council of Europe 2001, Petrakos and Psycharis 2004). More importantly, it inhibits the formation of local spaces of regulation. The restricted capacity of local authorities to actively finance and execute locally decided development plans limits interaction with interest groups, directing the latter's attention towards the respective gatekeepers at the national level (Chorianopoulos 2008). The centralised and hierarchical characteristics of Greek scalar architecture, framed in the distinct economic and socio-political context of the post-war period, were thus reflected in the underdeveloped relational traits noted locally. In order to address this hindrance to competitiveness, a range of state spatial restructuring initiatives were launched in the 1990s.

State Spatial Restructuring

The 1990s reform wave started with the transformation of the 54 administrative Prefectures into political bodies (1994), featuring a directly elected Prefect and Council. The duties of the new Prefectural units were oriented towards the provision of physical infrastructure and services, actions that municipalities faced pragmatic (financial and organisational) hindrances in assuming (GGN 1994). The new Prefectures provided a politically accountable planning platform, bridging municipal with regional authority actions. It was the municipal tier, however, that was the focal point of reforms.

Following an unsuccessful attempt at voluntary municipal amalgamations (1994), national authorities decided to compel change (Chorianopoulos 2012). The 1997 Local Authority Act aimed at creating new (larger) municipal units, the scale of which would be 'adequate enough' to support cost-efficient service delivery, and to influence local development prospects. As noted in Figure 8.3, the number of municipal units was reduced to 1,034 (as opposed to 5,775), while boundary redrawing was expected to act as a catalyst, fostering the involvement of a broader range of private and voluntary sector interests in policy-making (MoI 1997a).

Such changing geographies of regulation reflect contemporary EU spatial policy priorities (Giannakourou 2005). They do not, however, occur in a vacuum. Previous urban policy choices influence restructuring directions. Focusing on the 1997 territorial reform, for instance, new municipal boundaries were unexceptionally defined by the aggregate administrative limits of units that were joined. An attempt to identify and frame the 'never perfect' and 'ever changing' functional limits of a locality, it was stated, would only 'add confusion to reforms', raising opposition (MoI 1997b). Thus, the realisation of the 'preferable spatial scale or size in the new municipalities' became the main goal, overshadowing arguments supporting a boundary redrawing effort centring on 'grasping' local socio-economic dynamics. Subsequent reform stages present similar hesitancies in promoting the stated, otherwise, objective of local synergies. The 1999 'Integrated Urban Interventions' Act, for example, marks a shift from the dominant local authority preoccupation with infrastructure improvements, enabling the local state to embark on thematically broader partnership activities with interest groups. The Act, however, prohibits any partner organisation from initiating or running projects, assigning this responsibility solely to the local authority (Local Authorities Institute 2006). The absence of cooperation experiences at the local level, therefore, resulted in a timid reform endeavour that circumscribed the development of local relational dynamics.

The EU conceptualisation of urban intervention, therefore, differs from local realities. EU urban policy is informed by resurgent cities: areas that experienced de-industrialisation and managed to effectively restructure their economy based on concerted and locally-defined governance responses (Storper and Manville 2006). De-industrialisation was never a major urban concern in Greece. The key role of the service sector as an employment provider during urban centralisation shaped particular urban prospects. Because it followed the emergence of urbanised economies, industry is primarily located on the urban outskirts (Leontidou 1990). Its subsequent withdrawal, in turn, affected primarily detached parts of the urban fabric.[9] Moreover, limited local articulation of interests questions the effectiveness of a policy intervention aimed at triggering locally defined synergistic actions. It is in this context of centralised political-administrative structures and cautious

9 It is for this reason that the experience of Greek cities did not inform the de-industrialisation debate that dominated urban geography discussions in the 1980s. Greek inner cities were not part of this trend.

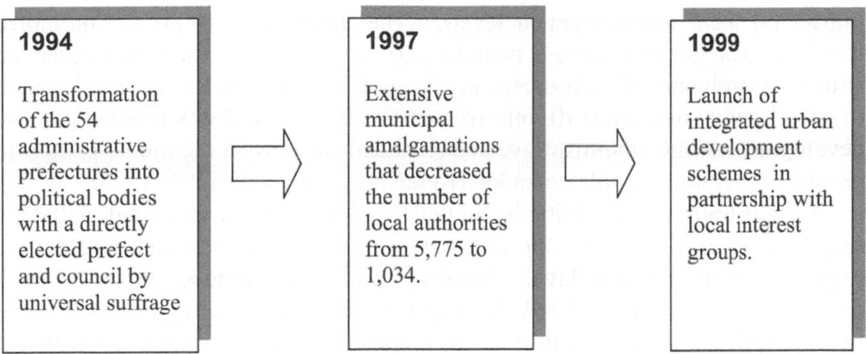

Figure 8.3 Key state spatial restructuring attempts (1990s)

Source: Local Authorities Institute (2006).

territorial experimentation that the impact of EU urban policy on Greek cities is going to be explored.

The case-studies are Keratsini-Drapetsona and Iraklion; selected on the basis of their record of URBAN Initiative participation, the programme that defined the current mode of EU urban intervention (CEC 2006b: 2). The fieldwork consisted of visits to the cities during the programming period and interviews with local authority personnel responsible for running URBAN, as well as with key private and voluntary sector partner organisations. Interviews were semi-structured, based on an outline of key questions that explored the following themes: a) the mode of local involvement in URBAN; b) the relationship between the local level, the national authorities, and the EU during implementation; c) the participatory traits of local interest groups (small and medium sized company associations, Chambers of Commerce, as well as community and educational bodies) in URBAN.

Analysis starts with a brief presentation of the socio-economic characteristics of the urban areas of interest, and focuses, subsequently, on local governance structures and responses.

Keratsini-Drapetsona

The Keratsini-Drapetsona URBAN programme centred on two neighbouring municipalities in the Piraeus metropolitan belt with 71,000 and 14,000 inhabitants, respectively. As Keratsini and Drapetsona are adjacent to the industrial docks of the Piraeus port, the employment structures of the area have been determined by the dominance of secondary sector activities (transport, storage and oil refineries). However, the restructuring of transport industries since the 1970s had a major impact on local unemployment levels.[10] The local URBAN project promoted five broad categories of targets including: a) support for industrial sectors facing structural problems; b) advancement of employment policies; c) development of social policy measures; d) improvements in the physical environment; and e) development of the administrative and technical capacity of the municipalities to promote programme implementation (Keratsini-Drapetsona 1996: 42–3).

Administration of the Initiative in Greece took the form of a single URBAN programme coordinated by the national authorities, with six sub-programmes implemented at the local level. There was no bidding process amongst Greek cities for participation in URBAN. The six URBAN sub-programmes and the corresponding local authorities[11] were nominated at the national level by the ministries responsible for the implementation of the second Community Support Framework (1994–2000). Information about URBAN was sent to Keratsini-Drapetsona local authorities by the Ministry of Environment and Planning (1994), and the two municipalities were asked to submit a combined Action Plan based on local policy priorities (Tsaousis 1997). The preparation of URBAN was assigned to the Development Corporation of Piraeus Municipalities (ANDIP), a municipal organisation founded in 1986 with its main areas of activity being the conduct of socio-economic studies and the implementation of various policies decided upon by the two local authorities. The local Development Corporation also had a central position in the administrative structures of the Initiative, portrayed diagrammatically in Figure 8.4.

At the national level, URBAN was regulated by a Monitoring Committee consisting of representatives from the respective Ministries,[12] the European Commission, national sectoral associations, and interest groups, as well as by local authority representatives from the six sub-programmes. The leading role in this Committee was exercised by the Ministerial tier, which had the responsibility to coordinate the local URBAN sub-programmes and guarantee the alignment of their targets with the overall Community Support Framework priorities (Ministry of Environment and Planning 1995). Two regulatory bodies were formed at the local level. First, the Local Steering Committee, comprising representatives from the Municipal Councils, the local development corporation, the corresponding National Ministries, and local interest groups. The local Steering Committee held monthly meetings and decided on projects and budget arrangements suggested by the second tier of the local URBAN administration, the Implementation Committee. In Keratsini-Drapetsona, the URBAN Implementation Committee

10 By the mid-1990s, activities related to transport, storage, oil refineries and power stations accounted for the employment of 46 per cent of the active labour force of the area, while unemployment had reached 21 per cent of the local labour force (Keratsini-Drapetsona 1996: 5, 7, 12).

11 Namely, Keratsini-Drapetsona, Volos, Syros, Patra, Peristeri and Thessaloniki.

12 Specifically, the Ministry of Environment and Planning and the Ministry of National Economy.

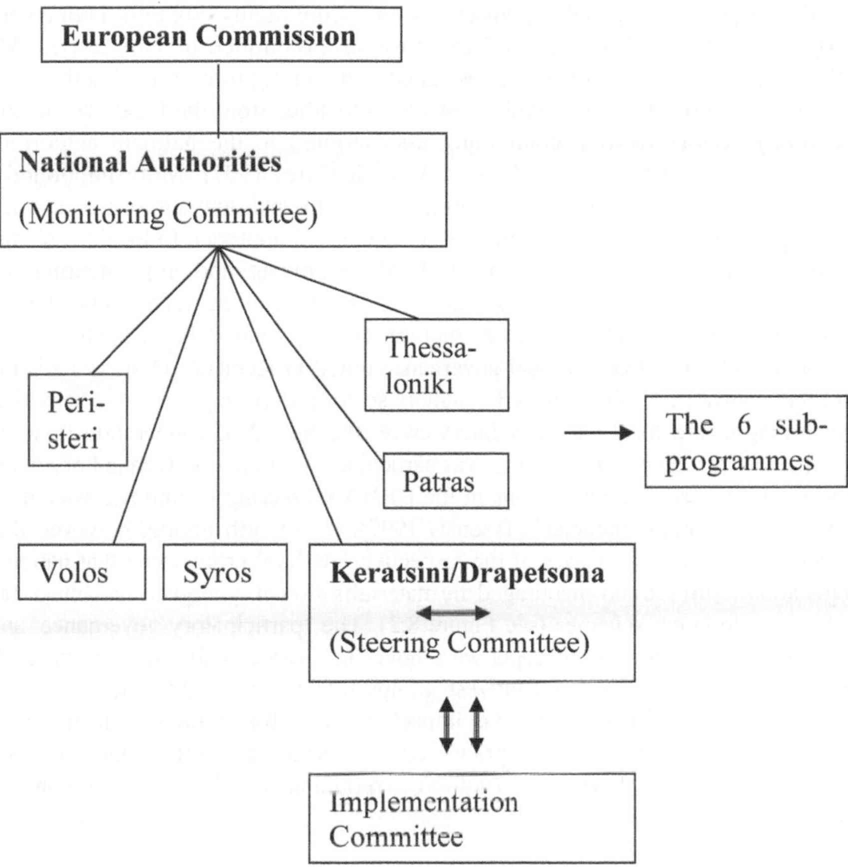

Figure 8.4 The administration of URBAN in Greece (1994–9)

was the local Development Corporation (Iggliz 1997, Ministry of Environment and Planning 1995, 1996: 8).

The centralised traits of the Greek administrative system had a strong impact from the very outset of URBAN, when national authorities controlled and selectively channelled local level access to the Initiative. In subsequent programme phases, the strong national grip on URBAN took various forms. The most prominent example comes from the policy areas of vocational training and small and medium sized enterprise (SME) support, which featured prominently in both the respective Regional Operational Programme and in the URBAN Initiative. In order to tackle the issue of overlapping targets, the URBAN Monitoring Committee of Greece issued a statement explaining, 'actions relevant to the development of SMEs will be organised centrally by the Ministry of National Economy', while, with regard to 'the development of vocational training programmes, these are to be implemented by the respective national organisations and Community Support Framework programmes assisted by the ESF' (Monitoring Committee of Greece-URBAN 1995: 1). In real terms, this policy assigned a mediating role to local authorities, which ended up forwarding applications for subsidies from the local private and voluntary sectors (SMEs, community associations) to the national authorities (Development Corporation of Piraeus Municipalities 1997). More importantly, the assumption of key parts of the programme by the national authorities arrested the opportunity for the development of an integrated approach to local problems. National authorities undertook the task of organising and implementing the majority of socio-economic regeneration measures, leaving local authorities to work on a number of physical infrastructure schemes funded by URBAN.

A second trait of Greek local governance noted in Keratsini-Drapetsona is the limited involvement of private and voluntary sector groups in programme activities. According to the local authority interviewee, the record of cooperation between local authorities and interest groups was particularly limited. Involvement of private and voluntary sector organisations in the URBAN Steering Committee was of an 'overall ceremonial character' (Tsaousis 1997). It is worth noting, however, that it was during the final phases of the respective URBAN programme that national legislation permitted and encouraged the materialisation of synergistic and integrated urban development schemes (see Figure 8.3). The 'participatory governance' and 'integrated interventions' concepts were novel in Greece at that time, putting ill-prepared local authorities and interest groups into the spotlight. The experience gained through URBAN, and the impact of the aforementioned institutional adaptations to EU spatial policy prioritisations, was more visible in the following programming period (URBAN II 2000–6), explored next in the city of Iraklion.

Iraklion

Iraklion is situated in the North-Western part of the island of Crete and also presents a case of de-industrialisation related intervention. The difference between the two URBAN case-studies, however, is that de-industrialisation in Iraklion affected primarily a relatively isolated part of the urban fabric, the old industrial port. This area apart, the city boasts a successful and multi-faceted restructuring record, suggesting an active local political level (Asprogerakas 2004). Examples range from the successful lobbying of the International Olympic Committee and the (exceptional) hosting in the city of a number of 'Athens 2004' Olympic Games events, also accompanied by significant investment in physical infrastructures. Longer term development strategies include the involvement of the city in two EU related local authority networks and the consequent participation in a number of Information and Communication Technologies (ICTs) projects.[13] The

13 Iraklion is member of the 'Eurocities' and 'Eurotowns' local authority networks, and participates in the respective 'Knowledge Society' and 'Capture' ICT projects.

extent to which the city is actively seeking to grasp development opportunities is indicated by the 2001 opening of the Brussels Office of the Iraklion Development Corporation, aiming to 'inform, negotiate, network and lobby directly in Brussels on behalf of the prefecture's municipalities' (Iniotaki 2004). It was in this context of awareness that funding opportunities that would facilitate intervention in the old industrial port were sought. Thus, when the Commission publicised the second phase of the URBAN Initiative, Iraklion did not let this opportunity slip away.

The Iraklion URBAN II (2000–6) programme focused on three adjacent districts in the North-West part of the city, with a population of 20,000: Agia Triada, Kaminia and Agios Minas. The port-related industrial activities that had dominated the area since the post-war years were severely affected by de-industrialisation, a trend that resulted in a deteriorating urban fabric and unemployment rates reaching 18 per cent of the local labour force; twice the city's average. Moreover, during the 1990s, comparatively lower housing rents attracted significant numbers of immigrant workers into the area, accentuating already noticeable social exclusion problems. Iraklion-URBAN II set three main goals for the area: a) physical and environmental regeneration that builds upon the local cultural and architectural heritage; b) job creation and SME development schemes; c) the establishment of a new social service network aimed at tackling social exclusion (Iraklion URBAN II 2001).

A number of changes were noted in the implementation of the second phase of URBAN, indicating a process of change in action. The most characteristic example comes from the realm of central–local relations, suggesting a trend towards a more open and decentralised institutional environment. In URBAN II, this was apparent right from the launch of the Initiative. The substantial budget that accompanies URBAN projects attracted the attention of Greek local authorities, which demanded a call for proposals open to all eligible cities. The Ministerial tier responded positively. In total 40 applications were submitted and three local authority schemes[14] were selected and approved, based on a range of qualitative and quantitative criteria (Manola 2004). In subsequent stages, two new single-purpose administrative units were set up at both national and local levels, the Steering Office and the URBAN II Office respectively, directly linking the two tiers and speeding up problem solving and project implementation. Despite decentralisation tendencies, however, past institutional choices left their own imprint on this process of change.

Leaving the two new administrative units aside, the political structure of URBAN II retained its centralised traits noted in the first phase of the Initiative (see Figure 8.3). The Ministerial level continued to hold decision-taking responsibility over all URBAN II sub-programmes, an arrangement that reflected the non-participation of local authorities in the Initiative's co-financing obligations. As a result, project funding was not in tune with the implementation process. Actions related to support for SMEs, for instance, were funded and launched in May 2007 – a year after the end of the official programme period – blocking any externalities

14 The cities of Iraklion, Perama and Komotini.

expected to materialise from the multi-faceted URBAN approach to intervention. In policy areas where action relied solely on local initiative, however, procedures were more efficient.

Local authorities in Iraklion managed to involve the majority of local stakeholders in URBAN,[15] promoting a well discussed and integrated action plan for the area in question. Interest group involvement was also reflected in the implementation stage, marking a stark contrast to past experience in the country. The University of Crete, for instance, organised and implemented the conversion of an old warehouse into the city's new natural history museum, administering directly 25.8 per cent of the URBAN budget. However, the role of other-than-public-sector interest groups in the implementation phase had to be adjusted to the national regulatory framework that guides such schemes. The local Chamber of Commerce and Industry, for instance, a private sector organisation, was not in a position to directly administer URBAN funds. Instead, it provided vocational training to local residents through the contracting-out of its services (Katharakis 2006).

Innovative policies were also noted in other areas. URBAN funds facilitated the creation of a new social service network in the old port. Activities assumed by the local authority include, amongst others, nursery and kindergarten facilities, a counselling unit for the young and the elderly, a home-care unit, a language training centre oriented to minority needs, a hostel for the homeless, and a drug rehabilitation centre (Iraklion URBAN II 2001). The fact that local authorities have limited room for influencing the investment layout of local income casts doubt on the financial viability of these schemes in the long run. The very development of this structure, however, is seen locally as a concrete claim for change; a lever that would enable the mitigation of rigid national controls that predetermine local spending (Kokori 2004).

Reflecting on the Research Findings

Key differences were noted in the responses of the two case-study cities to URBAN, underlined by the higher degree of effectiveness with which Iraklion approached and implemented the Initiative. Such differences reflect the particularity of local economic and socio-political circumstances. They are also suggestive, however, of the gradual impact of EU policies on Greek local governance. As discussed, during the time-span of URBAN (1994–2006) a number of state spatial restructuring reforms sought to strengthen the regulatory and developmental profile of the local political level, recognising its role in development prospects. It was in this experimentation period that Greek local authorities staked out their participation in URBAN II, opposing the controlling role of the national administration in EU funded programmes. It was also in this time period that Iraklion re-oriented its

15 Key examples include the local Chamber of Commerce and Industry, the Association of Trade Unions, the University, and the local residents' group.

development vision by opening up an office in Brussels, and structured a sound integrated intervention attempt in deprived port quarters. Institutional change, however, takes place in response to past experiences. The reliance of the local level on national finances was detected in both cities, arresting the emergence of local coalitions in the SME area. More characteristically, the limited involvement of private and voluntary sector groups was noted, reflecting the commanding role of the local political level in any integrated intervention attempt.

EU urban policy envisages a mode of local mobilisation that encompasses key stakeholders, creating a mediatory platform regulated by the local political level. In this ideal-typical neo-corporatist structure, participant policy views are expected to assume equal weight, different opinions to be consensually synthesised into a clear position (an action plan), with resources to be shared. The materialisation of synergy, in turn, spreads associated risks while simultaneously promoting a mutually binding involvement to restructuring. Underdeveloped governance experiences and rigid administrative hierarchies, however, rendered Greek localities unprepared for participating in such endeavours. Decision-taking verticality in Greece, as sketched in this chapter, was not conditioned by the particularities of local circumstances. It corresponded to the rules and structures of the national administrative framework, mitigating the restructuring impact of EU urban programmes. In this perspective, two recent developments suggest a changing setting. First, the upgraded role entrusted to cities in the current structural fund programming period (2007–13). Second, the latest state spatial restructuring attempt (2010) that re-orders central–local relations. The final section of this chapter explores these initiatives, commenting on the prospects and risks they engender.

Coda: The Challenges Ahead

Although a clear policy plan regarding urban areas was not available at the time of writing, the approach to cities presented in the National Strategic Reference Framework is challenging and innovative. The relevant document recognises urban social exclusion and unordered expansion as key socio-spatial problems in the country, while simultaneously portraying cities as engines of economic activity. Continuing, it categorises and ranks urban areas on the basis of a number of criteria,[16] to propose, in return, 17 cities as national growth nodes. It is around these cities that sustainable development targets are to be promoted in the current programming period (Ministry of Economy and Finance 2006). The incorporation

16 Urban category criteria are related to: population size and dynamics; economic performance and relevance of particular sectors to local economy; location with respect to major transport routes; presence of research facilities and administrative headquarters; and, degree of functional networking linkages with neighbouring cities, amongst others (Ministry of Economy and Finance 2006: 80).

of cities in a national spatial development perspective was not attempted in the past. That development alone stands as evidence of the upgraded importance of urban areas in Greek spatial planning. The spheres of activity that comprise the emerging national urban policy agenda revolve around the familiar ESDP (European Spatial Development Perspective) concepts of 'polycentrism' and 'urban–rural partnership', while equal attention is placed on the intra-urban environmental and socio-economic issues. What is not discussed in the NSRF documents, however, is the precise role that the local political level will be entrusted with in such an attempt. ESDP umbrella terms do not safeguard the active involvement of the local level in spatial policy, nor do they recognise the relational particularity of local governance. More importantly, by bridging different territorial and political-administrative levels, ESDP priorities re-engage national planning authorities in local development schemes. Responding to this quandary, national authorities initiated a state spatial restructuring reform (2010), oriented towards upgrading the role and functions of the local level in promoting development goals. The territorial and regulatory dimensions of the so-called 'Kallikratis Plan' were far reaching and extensive.

The new structures, launched in mid-2010, feature: a) the reduction in the number of municipalities from 1,034 to 325; b) the downright abolishment of the Prefectural tier of administration; and, c) the transformation of regions from an administrative arm of the state into political authorities, with a directly elected Head and Council. Municipalities were entrusted with enhanced financial autonomy and new responsibilities in urban planning, education and welfare policies. Regions retained their key role in NSRF processes. Moreover, they assumed metropolitan governance functions and the duties of the decentralised branches of national ministries in social and health policy. Seven 'Decentralised Administrations', in turn, a new territorial unit of the Greek state run by the national authorities, took over the task of monitoring the performance of the overall structure (see Chorianopoulos 2012). Enhanced participation, viewed as a key reform objective, was supported at all local authority levels (MoI 2010a).

In an effort to enhance collaboration, a 'Deliberation Committee' was established in municipalities and regions. This development-oriented governance platform involves key local stakeholders, while a quarter of its participants are randomly drafted citizens. At municipal level, the 'Committee's' advisory duties are facilitated by wide-ranging e-petition and e-governance provisions (MoI 2010b). Similarly, in Decentralised Administrations, the 'Council' engages the respective local authorities in policy-making and coordinates spatial planning policies in the area.

The latest state spatial restructuring attempt, therefore, perceives localities as 'action spaces' and local authorities as a political catalyst responsible for articulating place-based competitiveness responses (Schmitt-Egner 2002). The contextual risks associated with the competitiveness shift, however, have not been adequately discussed. Networking governance dynamics is not the necessary outcome of territorial and regulatory reforms. There is nothing inherent in larger scale territorial units or, for that matter, in local cooperation platforms in triggering competitiveness. Instead, potent competitiveness strategies build upon local relational assets; a meta-capacity that reflects untraded interdependencies between local actors (Storper 1997). This type of qualitative cooperation experience, however, is underdeveloped in Greek cities. In its absence, the quest for differential advantage is a venturesome undertaking, one that risks an increase in socio-spatial disparities (Turok 2009).

References

Andrikopoulou, E., Getimis, P. and Kafkalas, G. 1992. Local structures and spatial policies in Greece, in *Endogenous Development and Southern Europe*, edited by G. Garofoli. Aldershot: Avebury, 213–21.

Asprogerakas, E. 2004. The service sector as an inter-urban competition field: the role of medium-sized cities. *Geographies*, 8, 50–66.

Berg, L.V.D., Drewett, R., Klaassen, L.K., Rossi, A. and Vijverberg, C.H.T. 1982. *Urban Europe: A Study of Growth and Decline*. Oxford: Pergamon Press.

Camagni, R. 2002. On the concept of territorial competitiveness: sound or misleading? *Urban Studies*, 39(13), 2395–411.

CEC. 1992a. *Treaty on European Union*. Luxembourg: Office for Official Publications of the European Communities.

CEC. 1992b. *Urbanisation and the Functions of Cities in the European Community*. Luxembourg: Office for Official Publications of the European Communities.

CEC. 1993a. *Growth, Competitiveness, Employment. The Challenges and Ways Forward into the 21st Century – White Paper*. Luxembourg: Office for Official Publications of the European Communities.

CEC. 1993b. *Community Structural Funds 1994–99, Revised Regulations and Comments*. Luxembourg: Office for Official Publications of the European Communities.

CEC. 1994a. *Uruguay Round Implementing Legislation*. Brussels, COM(94) 414 final.

CEC. 1994b. *Competitiveness and Cohesion: Trends in the Regions. Fifth Periodic Report on the Social and Economic Situation and Development of the Regions in the Community*. Luxembourg: Office for Official Publications of the European Communities.

CEC. 1994c. *Community Initiative Concerning Urban Areas (URBAN)*. Brussels, COM(94) 61 final/2.

CEC. 1997. *Towards an Urban Agenda in the European Union*. Brussels, COM(97) 197 final.

CEC. 1998. *INFOREGIO NEWS: URBAN – Restoring Hope in Deprived Neighbourhoods*. Fact Sheet (15-11-1998 EN), Brussels: Directorate General for Regional Policy and Cohesion.

CEC. 2006a. *Cohesion Policy and Cities: The Urban Contribution to Growth and Jobs in the Regions*. Brussels, COM(2006) 385 final.

CEC. 2006b. Regulation (EC) No 1080/2006 on the European Regional Development Fund. *Official Journal of the EU* (31.7.2006), L 210, 1–11

CEC. 2007. *Cohesion Policy 2007–13. National Strategic Reference Frameworks*. Luxembourg: Office for Official Publications of the European Communities.

Cheshire, P.C. 1995. A new phase of urban development in Western Europe? The evidence for the 1980s. *Urban Studies*, 32(7), 1045–63.

Cheshire, P.C. 2006. Resurgent cities, urban myths and policy hubris: what we need to know. *Urban Studies*, 43(8), 1231–46.

Chlepas, N. 1997. *Local Government in Greece*. Athens: Sakkoulas.

Chorianopoulos, I. 2008. Institutional responses to EU challenges: attempting to articulate a local regulatory scale in Greece. *International Journal of Urban and Regional Research*, 32(2), 324–43.

Chorianopoulos, I. 2012. State spatial restructuring in Greece: forced rescaling, unresponsive localities. *European Urban and Regional Studies*, 19 (forthcoming).

Council of Europe. 1994. Definition and limits of the principle of subsidiarity. *Local and Regional Authorities in Europe, Study Series – Report No. 55*. Strasbourg: Council of Europe Press.

Council of Europe. 2001. *Structure and Operation of Local and Regional Democracy: Greece*. Strasbourg: Council of Europe Publishing.

DATAR. 2003. *Le villes Européennes: analyse comparative*. Montpellier: DATAR.

Development Corporation of Piraeus Municipalities. 1997. *Introduction for URBAN – Information*. Piraeus: ANDIP.

ESPON. 2006. *Territory Matters for Competitiveness and Cohesion. Facets of Regional Diversity and Potentials in Europe*. ESPON Synthesis Report III. Luxembourg: ESPON.

GGN. 1994. Prefectural self-government: legal status and responsibilities. *Greek Government Newspaper*, 90-A.

Giannakourou, G. 2005. Transforming spatial planning policy in the Mediterranean countries. *European Planning Studies*, 13(2), 319–31.

Hudson, R. and Lewis, J.R. 1984. Capital accumulation: the industrialisation of Southern Europe, in *Southern Europe Transformed: Political and Economic Change in Greece, Italy, Portugal and Spain*, edited by A.M. Williams. London: Harper & Row Publishers, 179–207.

Iggliz, V. 1997. Member of Drapetsona Municipal administration. Co-ordinator of Piraeus URBAN. Interview date: 18 April.

Iniotaki, H. 2004. Chief Executive of Iraklion Development Corporation, Brussels Office. Interview date: 1 October.

Ioakimidis, P.C. 1984. Greece: from military dictatorship to socialism, in *Southern Europe Transformed: Political and Economic Change in Greece, Italy, Portugal and Spain*, edited by A.M. Williams. London: Harper & Row Publishers, 33–60.

Iraklion URBAN II 2001. *Community Initiative URBAN II: Programme Iraklion. Economic and Social Regeneration of the Deteriorated Areas of the Western Iraklion Coast (Crete) 2001–2006*. Iraklion: Municipality of Iraklion.

Katharakis, M. 2006. Chairman of the Centres for Vocational Training of Iraklion Chamber of Commerce. Member of the URBAN Monitoring Committee. Interview date: 12 April.

Keratsini-Drapetsona. 1996. *Community Initiative URBAN*. Piraeus: ANDIP.

Kokori, H. 2004. Chief Executive of the Department of Social Policy, Iraklion Municipality. Interview date: 15 July.

Lalenis, K. and Liogkas, V. 2002. Reforming local administration in Greece to achieve decentralisation and effective space management: the failure of good intentions. *Discussion Paper*, 8.18, Department of Planning and Regional Development, University of Thessaly. Volos: UoT, 423–46.

Leontidou, L. 1990. *The Mediterranean City in Transition: Social Change and Urban Development*. Cambridge: Cambridge University Press.

Local Authorities Institute. 2006. *Urban Planning: Implementation Problems and Reform Proposals*. Athens: LAI.

Louri, H. 1988. Urban growth and productivity: the case of Greece. *Urban Studies*, 25, 433–8.

Manola, K. 2004. Chief Executive of Hellas URBAN Administrative Unit, Ministry of Environment, Planning and Public Works. Interview date: 16 July.

Ministry of Economy and Finance. 2006. *National Strategic Reference Framework 2007–2013*. Athens: MEF.

Ministry of Environment and Planning. 1995. *Community Initiative URBAN: Structure and Organisation*. Athens: MIP.

Ministry of Environment and Planning. 1996. *Action Plan of the URBAN Sub-programme of Keratsini-Drapetsona*. Athens: MIP.

MoI. 1997a. *Proposing to the Greek Parliament the Act on 'The Establishment of the First Local Authority Tier'*. Athens: Ministry of Internal Affairs. www.hellenicparliament.gr/UserFiles/2f026f42-950c-4efc-b950-340c4fb76a24/NOM_NOM_EE_2539_UB11.DOC [accessed: 22 May 2011].

MoI. 1997b. *The 'Ioannis Kapodistrias' Programme for the Reconstitution of the First Local Authority Tier*. Athens: Ministry of Internal Affairs. hwww.ypes.gr/kapodistrias/greek/kapo/program.htm [accessed: 22 May 2011].

MoI. 2010a. *Kallikratis Programme: Commentrary Report on the New Architecture of Local Self-government and Decentralised Administration*. Athens: Ministry of Interior, Decentralization and E-Government.

MoI. 2010b. *Kallikratis Programme*. Athens: Ministry of Interior, Decentralization and E-Government.

Monitoring Committee of Greece-URBAN. 1995. *Resolution of the First Meeting of the Monitoring Committee of Greece-URBAN, 04-12-1995*. Athens: MIP.

OECD. 2007. *OECD in Figures*. Paris: OECD Publications.

Petrakos, G. and Psycharis, I. 2004. *Regional Development in Greece*. Athens: Kritiki.

Schmitt-Egner P. 2002. The concept of 'region': theoretical and methodological notes on its reconstruction. *European Integration*, 24(3), 179–200.

Storper, M. 1997. *The Regional World*. New York: Guildford.

Storper, M. and Manville, M. 2006. Behaviour, preferences and cities: urban theory and urban resurgence. *Urban Studies*, 43(8), 1247–74.

Trojan C. 1998. *Agenda 2000 and the Role of Cities*. Speach to Eurocities Seminar, European Parliament Building. 26 May.

Tsaousis, K. 1997. Financial Administrator of the Development Corporation of the Drapetsona-Keratsini Municipalities. Senior Manager of Piraeus URBAN. Interview date: 24 April.

Tsoukalas, K. 1986. Employment and employees in the capital: opaqueness, questions and suggestions, in *Greece in Evolution*, edited by A. Manesis and K. Vergopoulos. Athens: Exantas, 163–240.

Turok, I. 2009. The distinctive city: pitfalls in the pursuit of differential advantage. *Environment and Planning A*, 41, 13–40.

UN. 2007. *World Population Prospects: The 2006 Revision*. New York: UN.

UN. 2008. *World Urbanisation Prospects: The 2007 Revision*. New York: UN.

Williams, A.M. 1984. *Southern Europe Transformed: Political and Economic Change in Greece, Italy, Portugal and Spain*. London: Harper & Row.

World Bank. 1984. *World Development Report: 1984*. Oxford: Oxford University Press.

Chapter 9
A City Hall for the Competitive City: Urban Management and Urban Governance in Slovenia

Irena Bačlija[1]

Introduction: Urban in Political Science

When preparing for the urban future one cannot overlook the grim picture of environmental and social detoriation in the cities. According to Yates (1977), the size and heterogeneity of a city prevents coherent planning and policy-making, making urban problems virtually impossible to resolve. However researchers have to be challenged and respond to emerging issues presented by the changing urban reality. As Stren (1996) believes, the research community has to disseminate its message more effectively to the wider policy and activist community, while maintaining credible levels of logical argument and scientific integrity. In order to respond to urban complexity, and even more, to propose a concept of addressing urban problems that would fit different urban environments, one has to ensure institutional complexity that would match. This task is too great for a single discipline, so studying the urban is progressively interdisciplinary, although real interdisciplinarity is according to Stren (1993) is impossible to achieve. The prevailing dominant disciplines (urbanism, urban sociology, architecture) should be challenged, giving voice to other disciplines, especially to political science and administrative science.

Within political science the specialised subfield of urban politics focuses on the cities and related subjects and processes. There are many theoretical political science approaches that should be mentioned for better understanding of urban politics. However urban politics has for many years drifted from political science both theoretically and methodologically (Sapotichne et al. 2007). The Elite Theories with the 'iron law of oligarchy' that found proof for economic and political elites, decided what decisions should be taken in the city (Domhoff 1970, Hunter 1953, Lynd and Lynd 1929). This could be dated back to Aristotle and Machiavelli, and their study on the role of the elites in the city. Next to test Elite

1 University of Ljubljana; Faculty of Social Sciences. Kardeljeva pl. 5; 1000 Ljubljana (Slovenia). irena.baclija@fdv.uni-lj.si.

Theories was Dahl (1967) who explored the theory's limitations through looking at the evolution of political leadership in New Haven (USA). He did discover oligarchic and elitist polity in the past, but in the twentieth century the leadership became increasingly pluralistic and diverse. Pragmatic coalition-building was policy sensitive, resulting in no one group in particular governing the city. In urban politics there was also a strong influence of Public Choice Theory that studies political behaviour in terms of preferences and responsiveness. This was especially applicable to local level (also cities) as voters could move to another local entity in search of maximising their welfare (Downs 1957). Public Choice Theory was criticised for generalising studies made in US cities, much like Urban Regime Theory (Hamel 1999, Stoker 1996, Stone 1989), that was proposed in 1980s and 1990s. Not alike proponents of Elitist Theory, which focuses on who governs, proponents of Urban Regime Theory try to answer the question on how to obtain appropriate capacities to deal with urban problems and development of urban community. Clarence Stone (1989) believes that the complex urban environment prevents any single political group taking over complete control, therefore regimes assembled from different actors are created, that have capacity for policy realisation. In this context, Urban Regime Theory is much like urban governance, a concept that has prevailed in the urban politics field (regarding city environment) in the last two decades. The main difference is that proponents of Urban Regime Theory wanted to prove that effective city management depends on the ability of city government to include other actors (mostly economic) into the policy-making process. In this context individual regimes are being built and their purpose is to solve individual problems. The weakness of Urban Regime Theory is that too much attention is placed on building partial policy networks and neglecting democracy and participation in the city. It legitimises great power that individual actors (especially economic actors) or groups of actors have on policy-making. In response, Urban Governance Theory emphasises democracy and transparency and devotes more attention to lack of legitimacy of these actors, and therefore the lack of taking over the responsibility for made decisions.

It is obvious that political science is mostly focusing on the question *who?* and on actions of political and economic elites or policy networks. Public administration on the other hand is dealing with the less popular aspect of the policy-making process – implementation. Therefore the main question is *how?* The *how* question was also partially addressed by urban politics studies, however only in connection to the type of the local government regime prevailing in some of the countries. Parker (2004) explains that since research on urban politics in some countries (e.g. Britain) was rooted in the study of public administration, some others (see Mollenkopf 1983) argued it lacked strong theoretical tradition and was often characterised by empiricism and comparative analyses. In this context cities are regarded as the places where local administration happens. In other countries (mostly in the United States) where cities enjoyed a higher level of autonomy, cities were observed more as polities. John (2001) illuminates marginalisation of studying local administration, with the notion that 'local government is perceived to administer things that others consider to be dull, such as bin emptying, street cleaning, drainage, and building quality; it is assumed the administration and decision-making of the these activities is similarly tedious'. But there is evidence that city administration is changing and is getting a more prominent position in developing strategies in neighbourhoods (Daemen and Schapp (2000) discovered evidence for this in Stockholm and Antwerp). Innovation and applying new methods of neighbourhood development are becoming the tasks of civil servants, with politicians operating only in the background.

However political science (and with it administrative science) has another *urban* concept on the agenda. That is urban management. Urban management is closely connected to the new role of local governments in the neoliberal era (Davey et al. 1996). As a concept, it has its roots in local government reform and geographical concepts of 'urban managerialism' in the 1970s, but it basically flourished as an institutionalised concept from the mid-1980s when it was championed by a number of key international donor agencies for the developing world (Jenkins 2000). Both concepts (urban governance and urban management) are commonly misinterpreted (a problem we will address later in the text), however used complementarily could offer a wholesome approach to resolving urban problems. It is indisputable that there is great demand for specialised tools which can help leaders juggle multiple urban challenges while working towards long-term solutions. Urban management and urban governance techniques and strategies have the potential to be developed into such a primer or playbook.

The goal of this chapter is first to present the relationship between urban management and urban governance, and the way these two concepts are commonly misinterpreted. Second, both concepts will be presented in Slovenia, a country that is defined by numerous adjectives, Mediterranean, South European (if you will) and post-communist being only a few of them. And in the final section, to present analyses of how urban management techniques are implemented in Slovenian cities, at the same time ranking them to EU cities.

Urban Governance

Discussing urban problems often leads to misusing certain terms that are commonly used. It is not uncommon that urban governance's and urban management's definitions overlap. In order to understand the difference, it is first necessary to present what urban governance is, and therefore what urban management is not. Note that management and governance are distinguishable. 'Management' refers to how officials execute the government's policies (i.e. delivering services and enforcing regulations), while 'governance' refers to more in the case of local government (Dijk 2006: 10). Rao (2007) seem to concur: 'cities are governed *and* managed.'

Let us first address what governance is. Governance, as distinct from government, refers to the relationship between civil society and the state, between the rulers and the ruled, the government and the governed (McCarney et al. 1995:

95). McCarney (1996: 4–6) believes that a good definition of governance is hard to find. She observes that the term *governance* is commonly misinterpreted as government or management. It should be noted that there are several ways to understand governance in structural terms. Pierre and Peters (2000: 14–22) present four different constructs: governance as hierarchies, governance as markets, governance as communities, and – as governance is most commonly understood – as networks.[2] The shift from the concept of government to the idea of governance is especially relevant in the local context, where democratic decision-making takes place in a smaller environment, making it possible for local authorities to respond to local demands more quickly. This conceptual shift also enables faster institutional changes and adaptations in the decision-making process. Urban governance cannot simply be equated with a more general concept of governance. McCarney (1996: 4–6) explains that when this definition is applied at a local level, a notion of urban governance helps to shift thinking away from state-centred perspectives which have predominantly focused on urban management to include elements which, in conventional terms, are often considered to be outside the public policy process. These include civic associations, private sector organisations, community groups and social movements, all of which in fact exert an impact on the features and development of urban centres.

Urban Governance vs. Urban Management

It is obvious that there is a relatively clear distinction between management and governance, therefore between urban management and urban governance. If, in the case of urban management, we refer to the officials executing the policies and, in the case of urban governance, we think of additional components (see Dijk 2006: 10). Even more illustrative we can say that if urban governance is a relation between government and the governed, then urban management is a relation between servers and the served (Reiss 1970).

Urban management is often described as an elusive concept (Mattingly 1994, Stren 1993) because academic and practitioner contributions to the debate have not converged, even within their own camps. Pahl (1975) originally pointed towards urban management or urban managerialism in his book entitled, *Whose City?* in which he suggested that urban resources are distributed by the managers or controllers of those resources. These so-called gatekeepers exercise a major constraint on the allocation of urban resources. The focus was on the allocation of scarce urban resources and the role urban managers played in the game of distribution. In this context, Williams (1978: 236) argued that 'urban managerialism is not a theory [or] even an agreed perspective. It is instead a framework for study'. He specifically observed the players in the process of resource distribution:

2 One of most familiar forms of contemporary governance is a policy network, which is not unlike the theories of urban regime (Hamel 1999, Stoker 1996, Stone 1989). This similarity is another reminder that governance is closely linked to urban space.

> There has been considerable debate as to whether urban managerialism should simply be concerned with the role of government officials (at both central and local levels) as mediators or whether it should encompass a whole range of actors in both public service and private enterprise who appear to act as controllers of resources sought by urban populations.

Leonard (1982: 10) seemed to concur, arguing that the origins of this managerialist thesis is the 'concern with the institutions and officials empowered to allocate resources and facilities'. As the 1980s progressed, the developing world experienced a shift in emphasis from the donor community. The provisions of mono-dimensional infrastructure schemes were being questioned as donors increasingly realised that such projects had inevitable and major consequences on other parts of the recipient's economic, social and environmental systems. Two changes were the gradual result. First, rather than deliver major engineering projects, donors moved towards a process of building institutional capacity and capabilities that allowed developing countries to provide and maintain their own infrastructure. This was the birth and growth of institutional development as a distinct intervention process in the developing world. Second, donors came to recognise the interconnections between various infrastructure projects, particularly in the urban sector. In that context, the academic examination of how to define urban management moved onto two different paths.

The first, and larger, group of authors considered urban management to be a process in which all interested parties (citizens, non-governmental organisations [NGOs], government, investors, etc.) took part in order to make the workings of a city meet their needs. Most commonly these definitions would be similar to Churchill et al.'s (1985: v):

> The term urban management is beginning to take on a new and richer meaning. It no longer refers only to systems of control but rather to sets of behavioral relationships, the process through which the myriad activities of the inhabitants interact with each other and with the governance of the city.

Similarly, Cheema (1993: 7) views urban management as a process of integrated and deliberative decision-making. Moreover, he considers that urban management is a holistic concept:

> It is aimed at strengthening the capacity of government and non-government organisations, to identify policy and program alternatives and to implement them with optimal results. The challenge of urban management is thus to respond effectively to the problems and issues of individual cities in order to enable them to perform their functions.

Bramezza (1996: 34) adds:

> Urban management can be defined as the co-ordinated development and execution of comprehensive strategies with the participation and involvement of all relevant urban actors, in order to identify, create and exploit potentials for the sustainable development of the city.

We should be cautious when understanding urban management as broadly as this. First, from the urban managerialist's point of view, urban management is not an integrated process of decision-making by all urban stakeholders and shareholders. Second, we already know a term for such a definition – urban governance. It could be that studies of urban governance are impinging on the urban management literature, which is why we will devote the next section specifically to the relationship between urban management and urban governance.

A second group of authors tried to explain the urban management concept as a slightly more implementation-oriented tool for tackling urban problems. To these authors, urban management is not merely the way governmental institutions and civil societies address the affairs of the city together. Rather, the concept of urban management has a practical connotation.

As Sharma (1989: 48) understands it:

> Urban management can be described as the set of activities which together shape and guide the social, physical and economic development of urban areas. The main concerns of urban management, then, would be intervention in these areas to promote economic development and well being, and to ensure necessary provision of essential services.

Davey (1993: xi) adds:

> urban management is concerned with the policies, plans, programs and practices that seek to insure that population growth is matched by access to basic infrastructure, shelter and employment. While such access will depend as much, if not more, on private initiatives and enterprise, these are critically affected by public sector policies and functions that only government can perform.

Rakodi (1991: 542) offers a similar view:

> Urban management aims to ensure that the components of the system are managed so that they make possible the daily functioning of a city which will both facilitate and encourage economic activity of all kinds and enable residents to meet their basic needs for shelter, access to utilities and services, and income generating opportunities.

As indicated by its name, urban management is first and foremost management. Chakrabarty (1998: 505) leads us to this conclusion when at first he defines management as the process of designing and maintaining an environment in which individuals, working together in groups, efficiently accomplish selected aims. Later he adds that urban management encompasses not only conventional business management, but also the additional tasks of managing the process of urbanisation, development and urban operations during which a number of urban enterprises have to work together to accomplish efficiently the diverse aims of the urban spatial units at various system levels and of the urban sector as a whole, involving multiple stakeholders.

According to Mattingly (1994: 201–5), urban management is 'public administration or growth management or organisational management'. He also tries to connect urban management to managerial processes, considering urban management to be 'taking sustained responsibility for actions to achieve particular objectives with regard to particular object'. He adds, 'this responsibility is to determine what needs to be done, to arrange that it be done, and then make certain that it is done for the city's development'.

Urban Government and *Urban* Governance in Slovenia

It is not coincidence that adjective urban in this heading is in italic. The reason for that is, it is hard to discuss *urban* in the country with only one real urban area. Beside Ljubljana, that Hall (1993) places in the category of 'smaller capitals and provincial capitals', that are commanding less extensive space in terms of population and GDP and are at the periphery of the European space (as well as Ljubljana also Dublin, Edinburgh, Lisbon, Helsinki, Stockholm and Bratislava), there is only one other area that can be marked as urban, although it only counts slightly over 110,000 inhabitants. When discussing urban government in Slovenia we are actually explaining municipal polity. Slovenian legislation de facto does not differ between urban and rural community. De jure there is a status of a city, that a town or municipality can obtain, with the decree of the national government. In order to be granted with that status several criteria have to be fulfilled; it should differ by size, economical structure and density of population from other settlements[3] and it should be inhabited by at least 3,000 citizens.[4] Today there are 68 *cities* with this status granted in Slovenia, however more commonly *larger*[5] urban areas are organised as *urban municipalities*. Urban municipality is a municipality that is densely populated, has at least 20,000 inhabitants; is place

3 However the Law on Local Self-Government (article 15a) does not state how it should differ.

4 This is not unlike other known criteria for the city. Most commonly a city is a settlement with at least 2,000 inhabitants (United Nations 2008), but usually also additional criteria has to be met. According to Cohen (1996: 226) also the number employed in the tertiary sector, population density, average income and other.

5 Note that by 'larger' the author refers to larger in the Slovenian context. Slovenia's largest urban municipality, Ljubljana, has 250,000 inhabitants and the smallest urban

for employment for 15,000 individuals; is the geographic, economic and cultural centre of its gravity area and fulfils other services criteria (high schools, hospitals, theatres, museums, etc.). In Slovenia we have 11 urban municipalities. Although legislation assigns urban municipalities with certain specific functions and duties, there is much room for further decentralisation or devolution of authority. We will address the authoritative weakness of Slovenian municipalities (also urban municipalities) later in the chapter, and for now continue with explaining the tasks of urban municipalities. In addition to general local matters, urban municipalities must also perform specific tasks that fall under national jurisdiction and that apply to the development of towns.

The basic needs of the population that the (non-urban) municipality is obligated to address include: primary education; primary health care; provision of essential utilities; postal and banking services; library; some form of public transportation; public space maintenance and use. In addition to these duties, urban municipalities also have to: regulate local public transportation; regulate public spaces and the construction of facilities; perform tasks in the area of geodetic services; administer a public network of primary, secondary, vocational and higher education institutions and libraries in their territories; ensure secondary public health service in their territories, including the administration of hospitals; provide a network of civil services; establish telecommunications centres and specialised information documentation centres, as well as local radio and television stations and press; support cultural activities (theatres, museums, archives) and sport and recreation facilities; administer all housing matters in accordance with the Housing Law, including maintenance of registers and contracts, monitoring of rents and issuance of construction permits and building inspections (Law on Local Self-Government).[6]

Municipal polity is determined by law and is the same in all Slovenian municipalities (therefore also in urban municipalities and municipalities granted with city status). It consists of three bodies: the municipal council, the mayor and the supervisory board. *The municipal council* is the highest decision-making body on all matters concerning the rights and duties of the municipality. It adopts general acts, approves the municipal budget and supervises the performance of the local administration and the mayor. The municipal council is comprised of between seven and 45 members proportional to the number of inhabitants in the municipality who are elected by citizens. *The mayor* represents the municipality, and is its legal representative. Slovenian system falls within Mouritzen and Svara's (2002) type of the strong-mayor form of government, although the mayor in the Slovenian system is not as strong as French or Spanish mayors. The mayor proposes the municipal budget, decrees and other acts within the jurisdiction of the council and is responsible for the implementation of council decisions. The mayor is the head of the municipal

---

municipality, Murska Sobota, almost 20,000. On average urban municipalities have 63,000 inhabitants and the average for all 211 municipalities (urban and rural) is 10,000.

6 Law on Local Self-Government (Official Gazette of the Republic of Slovenia, No. 72/93).

administration. *The supervisory board* regulates the management of municipal assets, ensures the purpose and efficiency of budgetary expenditures and monitors financial operations. Supervisory board members are appointed and dismissed by the council and may not be members of the council, municipal administrators, public employees or members of the management of budgetary organisations. Due to questionable independence (members of the supervisory board are appointed by the municipal council, the same council that adopts financial decisions) several calls have been made to scrap this mechanism altogether.

Although the main focus of this chapter is on urban management and the intertwining relationship between urban management and urban governance, urban governance itself should not be neglected. To stay within the framework of the general topic of *urban governance in Southern Europe* also urban governance in Slovenia has to be presented. As presented Slovenian urban governments' polity and politics are highly structured and predetermined by national legislation, which hampers cities' ability to adapt to market demands. The system of Slovenian local self-government is also highly financially centralised (local communities have little income from local taxes) that also prevent cities from investing in local projects. The best way to understand urban governance in this context is to observe how issues of democracy have affected the ways in which cities make choices. Cohen (1996: 16) outlines that municipal (also urban) leaders must learn how to create new resources to overcome small budgets, legislative constraints and other obstacles. They must become political entrepreneurs who garner resources and allies from the most unlikely places. We can assume that by *unlikely places* Cohen (1996) understood groups of stakeholders and shareholders. Since it is hard, if not impossible, to detect these pressure groups (that vary regarding policy area) we can asses capacity of urban governance by institutionalised ways of influencing policy-making in the city.

Slovenian legislation does not recognise shareholders as a legitimate pressure group on the local level, thus not stipulating mechanisms for cooperating in policy-making altogether. It is different with cooperation of shareholders. Forms of citizen participation (or groups of citizens' interest groups) at the local level are stipulated already in the Constitution of the Republic of Slovenia and subsequently the Law on Local Self-Government. Institutionalised forms of direct democracy are *citizens' assemblies*, *local referendums* and *people's initiatives*. A citizens' assembly is called, by the mayor, by the municipal council or the district council, or by 5 per cent of the voters in a municipality. In accordance with the law and the municipality's statutes, citizens discuss individual matters in the competence of the municipality and make proposals or pass decisions at these assemblies. Also various types of referendum can be held in local communities. The municipal council may, following a request by voters, call an advisory referendum on an issue of special importance to the local community or on acts concerning the municipality's affairs, with the exception of those concerning the budget, municipal taxes and other duties. The outcome of a referendum is binding for all municipal bodies until the expiry of their mandates. The issuance or cancellation of a general act within the competence of the municipal council or other municipal bodies may be enacted through people's initiatives by no less than 5 per cent of a municipality's voters. The body to which such an initiative is addressed must decide on the matter no later than within three months (Law on Local Self-Government).

Detecting whether these mechanisms of direct democracy are in common use is virtually impossible, since local governments are not obliged to report on this matter to any national body. The last research (taken in 2003!) showed that there were as few as seven local referendums, seven people's initiatives and 261 citizens' assemblies in a one-year period in all municipalities in the country. Although there are several different criteria on measuring the capacity of local (urban) governance, participation being only one of them, we have to some extent ignored other indicators at our peril. It should been noted that research on urban governance in Slovenia has so far not ben conducted.

Urban Management in Slovenia – Conclusions from the Research

The challenging part of the (re)conceptualisation of urban management is to define the substance of the concept. Based on the presented literature it can be concluded that the main task of urban management is the optimal function of city administration (therefore of the city itself). The next step would be the definition of the tools that enable the city administration to optimise functioning. When observing how individual city administrations deal with certain issues, categorisation of mechanisms can be proposed. These mechanisms are through institutional isomorphism in common use in cities globally. (Re)conceptualisation should be understood in this context and illustratively urban management should be understood as a basic rule of scale. Its chief concern should be to maintain a balance between the stakeholders (the citizens) and the shareholders (the investors), protecting and giving voice to citizens while at the same time providing opportunities for investors. The basic balance between social and economic development should be pursued. To attract investors to the city, a specialised labour force and infrastructure should be provided and the former is mostly attracted with job availability and high quality of life. So in order to attract investors, the labour force has to be attracted as well and vice versa. This centrifugal force works independently (Stren 1993), and the task of the urban management is to enable city administration to perform accordingly and to balance both demands.

The presumption was that urban management enables optimal function of city administration, which is oriented towards improvement of economic and social conditions in the city (Mumtaz and Wegelin 2001). Content of urban management should therefore present how a city (or a city's administration) enables these improved conditions. There is a wide range of good practices implying how one could successfully tackle varied and numerous urban problems (United Nations Human Settlements Programme 2004) and how to achieve optimal city development. A city's response could be different according to its predispositions (historical, legislative, macroeconomic) and environment (political, economic, administrative), however there is a way of managing that enables optimal adaptation to predispositions and environment characteristics. Therefore urban management should encompass all those dimensions that cover basic concepts of optimal city management. These dimensions should be paired to established administrative practices that have a positive effect on a city's successfulness (social and economic). In this context concepts of *city competitiveness, sustainable development, autonomy of urban manager, (user) participation* and *(sub)decentralisation* were used. 'Some other dimension could be included into urban management concept, as the concept is always changing and almost fluid, responding to the ever changing environment city administration has to function in' (Dijk 2006: 4).

In order to establish whether the urban management concept as proposed has been implemented in city administrations in the EU (also in Slovenia), we conducted a research titled 'Urban management in EU cities'. We have recoded urban managers' answers according to five urban management dimensions and correlated them (individually and as an Urban Management Index) with a number of independent variables. No particular school or theoretical approach guided the study. A rather open conceptual framework was established to identify institutional responses to urban problems. The study therefore examines how (with what tools, methods and approaches) city administrations dealt with the ever-expanding consequences of urbanisation. The core of our study is an interest in urban managers' perspective on these processes. In Table 9.1 there is a list of constituting dimensions of urban management and lists of indicators composing these dimensions.

Table 9.1 Indicators of urban management/Urban Management Index

Measures/dimensions	Hypotheses	Indicators
Urban decentralisation	• Decentralisation of the city creates more manageable entities. • Fiscal decentralisation of in the city results in more effective and more efficient public service provision. • Power decentralisation (e.g. decentralisation of policy-making) enables adoption of tailor-made policies.	• City is territorially decentralised (Hambleton 2004, Litvack et al. 1998, Stren 1993). • Individual city administration sectors are organised as dislocated units in subentities (Rondinelli 1990) • Subentities have elected representative political body (Bäck 2003). • Subentities are fiscally autonomous (Bäck 2003). • Decisions of subentities bodies are binding for city council/city administration (Bäck 2003).
Autonomous urban manager	• Cities with autonomous urban manager are economically more successful. • Cities with autonomous urban managers invest more into long-term projects.	• Urban management creates and/or adopts long-term city development strategy (Hill 2005). • Urban manager has more authority than the mayor/politician (De Long and Shleifer 1992, Hambleton et al. 2002, Leautier 2006, Rauch 1998, Svara 2003). • Appointment of professionals to executive positions through a selection process in which contracts are not tied to the duration of the political term (Borja 1996). • Governance implies more negotiation and coordination functions from urban manager (SOLACE 2004).

Measures/dimensions	Hypotheses	Indicators
Users' participation	• Including users into public services provision results in more effective and efficient public provision. • Including citizens into local policy-making results in adoption of tailor-made policies (therefore also more effective and efficient).	• Citizens are included in policy-making process (Beresford 2005). • City government consults with citizens (Beresford 2005). • Citizens are invited to comment (evaluate) on public services provision (Bäck et al. 2005: 130, Beresford 2005, Swindell and Kelly 2005).
City competitiveness	• City that is successfully competing on a global city market is economically more successful.	• City has a strategy for attracting (and/or keeping) tourists; investors; new citizens (Kotler et al. 1999). • City advertises its strategic advantages (to tourists; investors; new citizens) (Kotler et al. 1999). • Yearly plan of competitive indicators for the city is prepared (Alibegović and Kordej de Villa 2008). • There is an action plan for implementation of strategy for competitiveness (Phillips 1993). • Strategy for competitiveness is periodically evaluated (Phillips 1993).
Sustainable development	• Sustainable development enables long-term liveable environment, attractive for investors and for inhabitants.	• Limiting activities that pose threat to the environment and promoting environmentally friendly activities (Hardoy and Satterthwaite 1992). • Educating consumers and manufacturers on consequences of their activities (choices) (Hardoy and Satterthwaite 1992). • Using cost-benefit, SWAT, impact assessment tools for evaluating activities that pollute (OECD 2004). • Including experts and civil society into policy-making process (OECD 2004).

Note: districts, city quarters and neighbourhoods are considered subentities.

Measured indicators were recoded and merged into dimensions that are assessed on the scale from zero to five, zero being very low (e.g. urban manager is not autonomous; decentralisation has not occurred) and five being the highest (e.g. full sustainable development strategy in place; highest possible users' participation). At the last stage dimensions were summed into the Urban Management Index (scaled from zero to 25).

We have to be cautious when presenting results obtained with questionnaires. There are several limitations that pose a risk of unintentional generalisation, such as causality of some variables, respondents misunderstanding questions or terms, and others (see Armstrong and Lusk 1987, Heberlein and Baumgartner 1978, Singer et al. 1992). In Table 9.2 there are results for Slovenia's two biggest cities, Ljubljana and Maribor, and for the EU average.

Table 9.2 Urban management dimensions/index value in Slovenia and the EU

Measures/dimensions	**Value (0–5)**		
	Ljubljana	**Maribor**	**EU average (N=58)**
Urban decentralisation	2	3	2.2
Autonomous urban manager	2	4	2.7
Users participation	3	3	2.2
City competitiveness	5	0	2.8
Sustainable development	4	4	3.8
Sum/Index value (0–25)	16	13	13.3

Source: Bačlija (2010).

Firstly, both Slovenian cities are within the average value of the Urban Management Index, Ljubljana being slightly above the average. This is mostly due to the *City competitiveness* dimension value, which is assessed to be the highest in Ljubljana and non-existent in Maribor. Although Slovenian national legislation, as already mentioned, poses limitations to entrepreneurial activities of local entities, there is much room for creative manoeuvring. For example, an urban manager cannot attract highly skilled professionals into city administration since the local civil service is obliged to the same legislation as the civil service in general (also promotion, limitations in financial rewarding, limitations in working hours, etc.), but on the other hand tools like cost–benefit analyses or impact assessments can be implemented without limitations. In this context Slovenian cities don't have limitations in any of listed dimensions (especially if we look at the indicators). There are no limitations in urban decentralisation (which is determined in the city's statute), in awarding an urban manager with more autonomy (which is again determined in the city statute), with including citizens and users into policy-making and public provision processes, with creating and implementing competitiveness strategy and with achieving sustainable development. If proposed tools for enhancement of a city's success (economic, but also regarding better living conditions) are already possible and accessible, the question is why individual dimensions are scored so low (in Maribor city competitiveness and in Ljubljana urban managers' autonomy).

Conclusion

Economic globalisation, growingly differentiated and unstable markets, fiscal crises in light of increasing public expenditure are the elements of an uncertain future for cities. Some cities have preserved themselves as local societies; others lost their structure and have been for longer periods of time subjected to national and international strategies. In any case cities have to adapt to new circumstances through innovations in governing and administrating. In the chapter we have presented these shifts from governing and administrating to governance and management or more precisely to urban governance and urban management. Both have roots in the change of a development paradigm in the 1970s and 1980s (Brenner 1999, Harvey 1989, Keating 1998). Before that only two distinctive types of governmental functions were in use – production and consumption. The national level had the function of production and local communities took care of consumption with which they ensured continual manpower (Harding 2005, Harvey 1989, O'Connor 1973, Saunders 1986). Globalisation effects have caused the increasingly proactive role of cities that started planning their own development. Local authorities, even those from different socio-political environments, encountered the same challenge of economic development and their answers were similar innovative policies (Harding 2005, Mayer 1995, Porter 1998).

The conclusion reached in this chapter is that urban management is a reform of city administration and that its task is to create a much-needed balance between social and economic development. The two development fields have a fragile coexistence. In order to attract investors, we have to provide a suitable labour force and the labour force can only be attracted with jobs and quality of life (infrastructure, housing, services, etc.). A balance can be established by implementing the five dimensions of urban management: city decentralisation; user participation; autonomous management; sustainable development; and city competitiveness. These dimensions act as a fluid contextualisation of the concept, since new dimensions are always possible, depending primarily on the broader socio-economic and regulatory framework.

References

Alibegović, D.J. and Kordej de Villa, Ž. 2008. The role of urban indicators in city management: a proposal for Croatian cities. *Transition Studies Review*, 15(1), 63–80.

Armstrong, J.S. and Lusk, E.J. 1987. Return postage in mail surveys: a meta-analysis. *Public Opinion Quarterly*, 51(3), 233–48.

Bäck, H. 2003. *The Partified City: Elite Political Culture in Sweden's Two Biggest Cities*. Göteborg: School of Public Administration.

Bäck, H., Gjelstrup, G., Helgesen, M., Johansson, F. and Klausen, J.E. 2005. *Urban Political Decentralisation: Six Scandinavian Cities*. Wiesbaden: VS Verlag.

Bačlija, I. 2010. *Urban Management: Concept, Dimensions and Tools*. Ljubljana: FSS Publishing House.

Beresford, P. 2005. Service user: regressive or liberatory terminology? *Disability & Society*, 20(4), 469–77.

Borja, J. 1996. Cities: new roles and forms of governing, in *Preparing for the Urban Future: Global Pressures and Local Forces*, edited by M.A. Cohen, B.A. Ruble, J.S. Tulchin and A.M. Garland. Washington, DC: The Woodrow Wilson Center Press, 70–89.

Bramezza, I. 1996. *The Competitiveness of the European City and the Role of Urban Management in Improving City's Performance*. The Hague: CIP-Data Koninklijke Bibliotheek.

Brenner, N. 1999. Globalisation as reterritorialisation: the re-scaling of urban governance in the European Union. *Urban Studies*, 36(3), 431–51.

Chakrabarty, B.K. 1998. Urban management and optimizing urban development models. *Habitat*, 22(4), 503–22.

Cheema, S.G. 1993. The challenge of urban management: some issues, in *Urban Management Policies and Innovations in Developing Countries,* edited by S.G. Cheema and S.E. Ward. London: Praeger Westport, 1–17.

Churchill, A., Lea, J.P. and Courtney, J.M. eds. 1985. *Cities in Conflict: Studies in the Planning and Management of Asian Cities*. Washington: The World Bank.

Cohen, M.A. 1996. The hypothesis of urban convergence: are cities in the North and South becoming more alike in an age of globalization?, in *Preparing for the Urban Future: Global Pressures and Local Forces*, edited by M.A. Cohen, B.A. Ruble, J.S. Tulchin and A.M. Garland. Washington, DC: The Woodrow Wilson Center Press, 31–66.

Daemen, H. and Schaap, L. 2000. *Citizen and City: Developments in Fifteen Local Democracies in Europe*. Delft: Eburon.

Dahl, R.A. 1967. The city in the future of democracy. *The American Political Science Review*, 61(4), 953–70.

Davey, K. 1993. *Elements of Urban Management*. Washington, DC: The World Bank.

Davey, K., Batley, R., Devas, N., Norris, M. and Pasteur, D. 1996. *Urban Management: The Challenge of Growth*. Aldershot: Ashgate.

De Long, B. and Shleifer, A. 1992. *Princes and Merchants: European City Growth before the Industrial Revolution*. Cambridge, MA: National Bureau of Economic Research at Harvard University.

Dijk, M.P. van. 2006. *Managing Cities in Developing Countries: The Theory and Practice of Urban Management*. Cheltenham: Edward Elgar Publishing.

Domhoff, G.W. 1970. *Who Rules America?* New York: McGraw-Hill.

Downs, A. 1957. *An Economic Theory of Democracy*. New York: Harper & Row.

Hall, P. 1993. Forces shaping urban Europe. *Urban Studies*, 30(6), 883–898.

Hambleton, R. 2004. *Beyond New Public Management: City Leadership, Democratic Renewal and the Politics of Place*. City Futures International Conference, Chicago, USA, 8–10 July.

Hambleton, R., Savitch, H.V. and Stewart, M. eds. 2002. *Globalism and Local Democracy*. Houndmills: Palgrave.

Hamel, G. 1999. Bringing Silicon Valley inside. *Harvard Business Review*, 71(5), 70–87.

Harding, A. 2005. Governance and social-economic change in cities, in *Changing Cities: Rethinking Urban Competitiveness, Cohesion and Governance*, edited by I. Turok. London: Routledge, 62–77.

Hardoy, J.E. and Satterthwaite, D. 1992. *Environmental Problems in Third World Cities: An Agenda for the Poor and the Planet*. London: International Institute for Environment and Development.

Harvey, D. 1989. *The Urban Experience*. Oxford: Blackwell.

Heberlein, T.A. and Baumgartner, R. 1978. Factors affecting response rates to mailed surveys: a quantitative analysis of the published literature. *American Sociological Review*, 43(4), 447–62.

Hill, H. 2005. *Urban Governance and Local Democracy*. EGPA Conference 'Reforming the public sector. What about the citizens', Bern, Switzerland, 31 August.

Hunter, F. 1953. *Community Power Structure: A Study of Decision Makers*. Chapel Hill: University of North Carolina Press.

Jenkins, P. 2000. Urban management, urban poverty and urban governance: planning and land management in Maputo. *Environment and Urbanization*, 12(1), 137–52.

John, P. 2001. *Local Governance in Western Europe*. London: Sage.

Keating, M. 1998. *The New Regionalism in Western Europe: Territorial Restructuring and Political Change*. Cheltenham: Edward Elgar.

Kotler, P., Asplund C., Rein, I. and Heider, D. 1999. *Marketing Places Europe*. London: Pearson Education Limited.

Leautier, F. 2006. *Cities in Globalizing World: Governance, Performance and Sustainability*. Washington, DC: The World Bank.

Leonard, S. 1982. Urban managerialism: a period of transition? *Progress in Human Geography*, 6(2), 190–215.

Litvack, J., Ahmad, J. and Bird, R. 1998. *Rethinking Decentralization at the World Bank*. Washington, DC: The World Bank.

Lynd, R. and Lynd, H. 1937. *Middletown in Transition: A Study in Cultural Conflicts*. New York: Harcourt.

Mattingly, M. 1994. Meaning of urban management. *Cities*, 11(3), 201–5.

Mayer, M. 1995. Post-fordist city politics, in *Post-Fordism: A Reader*, edited by A. Amin. Oxford: Blackwell, 316–37.

McCarney, P. 1996. *Cities and Governance: New Directions in Latin America, Asia and Africa*. Toronto: University of Toronto.

McCarney, P., Halfani, M. and Rodríguez, A. 1995. Towards understanding of governance: the emergence of an idea and its implications for urban research in developing countries, in *Perspectives on the City*, edited by R. Stren and J. Bell. Toronto: Centre for Urban and Community Studies, 91–141.

Mollenkopf, J. 1983. *The Contested City*. Princeton: Princeton University Press.

Mouritzen, P.E. and Svara, H.J. 2002. *Leadership at the Apex: Politicians and Administrators in Western Local Governments*. Pittsburgh: University of Pittsburgh Press.

Mumtaz, B. and Wegelin, E. 2001. *Guiding Cities*. Washington, DC: The World Bank.

O'Connor, J. 1973. *The Fiscal Crisis of the State*. New York: St. Martin's Press.

OECD. 2004. *Policies to Enhance Sustainable Development*. Paris: OECD.

Pahl, R.E. 1975. *Whose City? And Further Essays on Urban Society*. Baltimore: Penguin Books.

Parker, S. 2004. *Urban Theory and the Urban Experience: Encountering the City*. London: Routledge.

Phillips, A. 1993. The growth of the conurbation, in *The Potteries: Continuity and Change in a Staffordshire Conurbation*, edited by A. Phillips. Stroud: Alan Sutton, 107–29.

Pierre, J. and Peters, B.G. 2000. *Governance, Politics and the State*. New York: St. Martin's Press.

Porter, M.E. 1998. *On Competition*. Boston: A Harvard Business Review Book.

Rakodi, C. ed. 1991. *The Urban Challenge in Africa*. Tokyo: United Nations University Press.

Rao, N. 2007. *Cities in Transition: Growth, Change and Governance in Six Metropolitan Areas*. London: Routledge.

Rauch, W. 1998. Problems of decision making for a sustainable development. *Water Science Technology*, 38(11), 31–9.

Reiss, A. 1970. The services and the served in service. *Urban Affairs Annual Review*, 4(3), 561–76.

Rondinelli, D.A. 1990. *Decentralizing Urban Development Programs: A Framework for Analyzing Policy Options*. Washington: USAID Office of Housing.

Sapotichne, J., Jones, D. and Wolfe, M. 2007. Is urban politics a black hole? Analyzing the boundary between political science and urban politics. *Urban Affairs Review*, 43(3), 76–106.

Saunders, P. 1986. *Social Theory and the Urban Question*. London: Routledge.

Sharma, K.S. 1989. Municipal management. *Urban Affairs Quarterly – India*, 21(4), 47–53.

Singer, E., Hippler, H.J. and Schwartz, N. 1992. Confidentiality assurances in surveys: reassurance or threat? *International Journal of Public Opinion Research*, 4(3), 256–68.

SOLACE. 2004. *Solace Member Survey 2004.* [Online]. Available at: www.solace.org.uk/library_documents/SOLACE_Summary_Survey.pdf [accessed: 2 November 2008].

Stoker, G. 1996. Redefining local democracy, in *Local Democracy and Local Government*, edited by L. Pratchett and D. Wilson. Basingstoke: Macmillan, 188–209.

Stone, C.N. 1989. *Regime Politics*. Lawrence: University Press of Kansas.

Stren, R.E. 1993. Urban management in development assistance: an elusive concept. *Cities*, 10(2), 125–39.

Stren, R.E. 1996. Administration of urban services, in *Making Cities Work: The Role of Local Authorities in the Urban Environment*, edited by R. Gilbert, D. Stevenson, H. Girardet and R. Stren. London: Earthscan Publications, 62–112.

Svara, J.H. 2003. Effective mayoral leadership in council-manager cities: reassessing the facilitative model. *National Civic Review*, 92(2), 157–72.

Swindell, D. and Kelly, J. 2005. Performance measurement versus city service satisfaction: intra-city variations in quality? *Social Science Quarterly*, 86(3), 704–24.

United Nations. 2008. *Demographic Yearbook 2006*. New York: United Nations.

United Nations Human Settlements Programme. 2004. *The State of the World's Cities*. Nairobi: United Nations.

Williams, P. 1978. Urban managerialism: a concept of relevance? *Area*, 10(3), 236–40.

Yates, D. 1977. *The Ungovernable City: The Politics of Urban Problems and Policy Making*. Cambridge, MA: MIT Press.

Chapter 10
Urban Governance in Istanbul[1]

Nil Uzun[2]

Introduction

In parallel with the effects of globalisation on policy issues, the topic of governance has been extensively investigated since the 1980s. In broad terms, governance represents a shift in the roles of formal government structures and contemporary agencies. There is also a change in the distribution of responsibilities between public, private, voluntary and household groups. Increasing fragmentation of responsibilities in the urban arena have increased the importance of new institutional relations and the policy process of different constituents and agencies within the existing national and local levels. The effects of this fragmentation and the rescaling process of the state are reflected as networked forms of governance. The relations between continental, national, regional and local governments together establish a new form of politics. The interaction of economic and institutional factors, mediated through political, cultural and other contextual means have been influencing the changing governance structure of cities and regions. As a result, the relationship between urban development and policy has become more complicated. A satisfactory urban governance model that can adequately represent all cases has not been developed.

The effects of globalisation have also been observed in Turkey in the last three decades and, therefore, there is an ongoing restructuring process. The government structure is being transformed towards governance. Nevertheless, the national government still has a significant influence on metropolitan development through policy-making. There is also a multi-level, multi-institution structure throughout a wide range of organisations. The goal of this chapter is to present a critical evaluation of urban governance. The thesis is that the generally defined concept of urban governance cannot be efficient in explaining the urban management model in countries on the periphery. The case of Istanbul is provided as an example for this discussion.

---

1 An early version of this text was published in 2010 in *Análise Social. Revista do Instituto de Ciências Sociais da Universidade de Lisboa*, 197, 757–70.

2 Middle East Technical University; Department of City and Regional Planning. 06531 Ankara (Turkey). duruoz@metu.edu.tr.

Defining Urban Governance

The term *governance* has its theoretical roots in many academic fields including institutional economics, international relations, development studies, political science and public administration. According to Schimitter 'Governance is a method/mechanism for dealing with a broad range of problems/conflicts in which actors regularly arrive at mutually satisfactory and binding decisions by negotiating with each other and cooperating in the implementation of these decisions' (Schimitter 2002: 53 cited in Haus and Heinelt 2005: 19). Considering several different usages of the term governance, Rhodes (1996) lists the common characteristics of governance as interdependence between organisations; continuous interaction between network members; game-like interactions; and a significant degree of autonomy from the state. When compared with government, 'Governance is … a more encompassing phenomenon than government. It embraces governmental institutions, but it also subsumes informal, non-governmental mechanisms whereby those persons and organisations with this purview move ahead, satisfy their needs and fulfill their wants' (Rosenau 1995: 4). Governance also refers to coordination of various interdependent activities (Jessop 1998).

On the other hand, the extensive literature on urban governance shows that it is not a newly emerging issue but it is very obvious that its context has changed in the last three decades. In the first half of the nineteenth century, there was a virtual governance structure based on utilitarianism. The assumption was that maximum public benefit would arise from free market forces. Between 1850 and 1910, *municipal socialism* was introduced by social leaders in response to epidemics, urban disorder and city congestion. Between 1910 and 1940, the Great Depression in the United States shifted public opinion in favour of a permanent and more fundamental government role in shaping many aspects of social life and well-being. From the following years until 1975, various methods of urban government generated large, vertically segregated bureaucracies of professional administrators geared towards managing the cities and their environment (Knox and Pinch 2000).

Since the 1980s, radical economic transformation at the national and international scale has set in motion a metropolitan restructuring process and governance has become the main trend. New economic scales need new methods of governance in order to negotiate new economic and territorial identities in the urban areas. As well as these, due to changes in the economic structure, city competitiveness has become the major driving force, and the concept of the entrepreneurial city has been replacing the concept of the managerial city and, therefore, growth coalitions are gaining importance. There has been a shift from government to governance. This is equivalent to focusing less on the institutions of government and more on the processes through which government institutions interact with civil society as well as the consequences of this mutual influence between the state and the society. Through this transition, the functions of formal government structures and contemporary agencies have also shifted. There is a new allocation scheme for the responsibilities of public, private, voluntary and household groups. The main problem in the urban arena has become the new institutional relations and policy processes of various constituents and agencies at the national and local level. Under the influence of globalisation, there are also changes in urban governance towards competition-oriented, innovation-oriented policies and new bargaining systems (Davoudi 1994, Jessop 1995, Knox and Pinch 2000, Mayer 1994).

According to Brenner (2004: 447) 'Urban governance … represents an essential institutional scaffolding upon which the national and subnational geographies of state regulation are configured as well as one of the major politicoinstitutional mechanisms through which those geographies are currently being reworked'. Pierre (1998) states three trends in urban governance. First, as a focus for proactive development strategies local politics have gained in importance. Second, in support of economic development, there is an increasing mobilisation of local politics which is observed in the local economic interventionism and the reorganisation of public services. As a final point, new bargaining systems and new forms of public–private partnerships are being redefined due to the expansion of the sphere of local political action.

Urban governance is based on the explicit representation and coordination of functional interests active at the local level. There is a cooperative style of policy-making and the local authority moderates or initiates cooperation instead of giving orders. In these new forms of urban governance, the actors in economic development and technological modernisation programmes are business associations, chambers of commerce, local companies, banks, research institutes, universities and unions, and the expanded sphere of local political action includes additional sets of actors such as welfare associations, churches, unions, grassroots initiatives and community organisations (Mayer 1994).

In the last three decades, parallel to the changing structure of the city, public–private partnerships in urban renewal and urban development projects have been gaining importance. There is an upgrading process in the central business districts and very often old industrial sites are being reintegrated into the city. The local governments aim to develop new attractive urban regeneration projects, leading to new partnerships with large investors, developers and consortia of private firms. There is a deal between public and private participants. Local governments provide the subsidy, power and necessary modifications in government regulations. While private partners meet certain project goals, take on later management tasks, and share project returns with the local authority. This type of collaboration provides the local authority with the ability to attract more financial resources to urban development and increase their effectiveness in achieving development goals, and the private sector is able to find attractive ways of expanding their activities and the ability to access the real-estate market (Carley 2000, Marshall 2000, Pierre 1998).

Since the 1980s, Turkey has also been experiencing the effects of globalisation. There is an ongoing restructuring process, within which government structures are moving towards governance. Together with this, urban redevelopment projects are also gaining importance in many metropolitan cities and in Istanbul. In the following sections, the changing government structure in Turkey and urban governance practices in Istanbul are subject to review.

Urban Government Structure in Turkey

The urban government structure of Turkey has been changing ever since the foundation of the Republic in 1923. Turkey subsequently adopted a single-party regime with a rather authoritarian structure. This was unique in that it tried to achieve both pluralism and revolutionalism and was not totalitarian. As a result of the authoritarian structure, municipalities were seen as the main providers of various services such as urban planning and health care. Many other fundamental problems, such as income resources for municipalities, were ignored. Due to its centralist approach, the state had the municipalities under its control. The first municipal system in the Republican period was established after Ankara was declared the capital city with the founding of the Prefecture of Ankara. After law number 417 was enacted on 16 February 1924, urban physical development started to accelerate in Ankara. This law had an important role in the development of other Turkish cities and their municipalities as well. Between 1930 and 1944, the municipalism approach became stronger with the passing of various new laws. Furthermore, Municipality Law number 1580 was approved in 1930. This law annulled all previous laws in effect since 1930. Afterwards changes related to income, municipality management and the organisation of planning functions, and some specific responsibilities like health and police services, occurred (Hamamcı 1990, Tekeli 1978).

The 15-year period following the 1945 transition to a multi-party era proved important since it was believed that local governments were basic elements in the democratic structure and that these must be ruled by governors and committees elected by the population; that the needs of the people had to be met at this level and the unique needs of provinces had to be observed and finally there had to be public participation. It was stated that the local governments must have enough power to meet these duties. In this period, the mayor and the province governor were separate. The state still had control over the municipalities. Following the 1960 coup, the 1961 Constitution stated that the administration should be a whole, which meant that the centre and local units should work together and form a whole. The Constitution considered local governments as Province, Municipality and Village and these were classified as public legal personalities whose decision units were elected by the public and who would meet common local needs. Hence, no structural changes to local government units were brought in by the 1961 Constitution. In the Constitution, the framework for *independent and powerful* local government was set out in principle, with application left to legislation. However, these frameworks were never completely developed and local governments were never as independent and powerful as stated (Altaban 1990).

In the 1973–80 period, social democratic parties had influence in local governments. In the 1973 elections, although the state was governed by conservative ideologies, city governments were under the control of social democrats. As a result, conflict between local governments and the central government occurred for the first time. The central government tried to restrict local government access to political benefits. Between 1973 and 1977, municipalities started to produce the goods and services they needed for providing the services themselves. Financing also represented a challenge to municipalities. In order to resolve this, municipalities started to find ways of creating resources themselves and they formed Municipal Unions campaigning against central government regardless of the political parties they belonged to.

The most recent period, which caused radical changes in the Turkish economy and the urban government system, started with the introduction of the privatisation model in the 1980s, consistent with the globalisation processes worldwide. The spatial impacts of these models along with the political priorities of the ruling parties and technological developments were reflected in the development of Turkish cities. Under the impact of the world economic crises and subsequent globalisation processes, the economic development model based on import substitution was replaced with a model encouraging export under the leadership of the private sector. This made it necessary to step up interaction with the world market and to integrate with the global system. These changes brought the need for new organisations in order to meet the requirements of the new economic structure. Thus, the business and service sectors began to gain importance in Turkey. Moreover, the state stopped investing in various industries and started to privatise those it owned by selling off factories. As a result, production and industrial investments increasingly experienced free market economy conditions (Kepenek 1999).

As a result of these changes in the government structure through to the 1980s, Turkey experienced a complex and multi-aspect socio-economic and political environment. In the last three decades, the impact of democratisation processes, external factors closely related with globalisation, developments in relations with international organisations such as the European Union and local factors have led to the restructuring of Turkey's government bodies in accordance with the governance concept (Kovancı-Shehrin 2005).

In 1984, with the introduction of new legislation (law number 3030), related to the government of larger metropolitan areas, important changes were made. First of all, instead of a single metropolitan administration, which was proposed in the military era, the Metropolitan City concept was brought in. According to this new organisation, metropolitan areas with more than one district in their borders were considered as Metropolitan Municipalities (MMs). Secondly, the financial resources of local governments were increased. Thirdly, the authority for making urban physical development plans was given to municipalities. Authority over development issues was transferred from central to local governments. This was a first step towards decentralisation. Relative independence was granted to municipalities, and financial resources were provided by the central government to MMs. This independence was reflected in municipal mega-projects. However, the central government still retained much control over MMs (Gülöksüz and Tekeli 1990).

The decentralisation process experienced in this period sought to transfer power from central to local government. Conflicts between central government and local governments, scarcity of financial resources, the lack of institutional and personal capacities, overlapping functionalities, and the lack of participatory and consultative mechanisms have been critical problems in this decentralisation process.

Along with these developments, the HABITAT II Conference, held in Istanbul in 1996, represented a critical external factor signifying the governance debate in Turkey. In this conference, governance issues were discussed to a greater extent within the socio-political agenda. The main themes of the conference regarding governance were changing state–society relationships, the increasing importance of civil society developments, the inefficiencies of representative democracy, and the need for participatory democracy (Kovancı-Shehrin 2005).

The changes mentioned above have been effective in most of the metropolitan cities of Turkey, but Istanbul is the only city where the effects of globalisation can be observed in many dimensions. Istanbul became an important centre for manufacturing and a connecting point in the world system. Today, Istanbul is the most dynamic city in Turkey. Although it is not the capital city, it is the centre of finance, transportation routes and industrial activity and has an important role in the integration process of the world system. Istanbul may correspondingly be considered as an example against which to test the relevance of generally accepted characteristics of urban governance.

Urban Governance in Istanbul

Istanbul, the largest metropolis of Turkey, has been an economic hub throughout history due to its advantageous location. By connecting Europe and Asia, it has enabled the provision of raw materials from Anatolia and the marketing of goods to the world. As the only city in the world being a capital during two consecutive empires and with the remnants of many civilisations, Istanbul has maintained its importance as an economic, social and cultural centre through the centuries. It is now the largest city in Turkey with respect to population size (over 13 million in 2010), the scale of economic activity, and the extent of its hinterland.

Since the foundation of the Turkish Republic, the problems and peculiarities of rapid urbanisation have always been present in Istanbul. The acceptance of the privatisation model in Turkey under the impact of globalisation also marked a turning point in urban change in Istanbul. As the most important metropolitan centre of the country, Istanbul became the foremost candidate to obtain a location for itself in the network of global cities. In this context, the economic base of the city started to change. It proved an attractive centre for foreign investors and the hub of an international communications network. This new, open economic connection to capitalist systems around the world stimulated local capital and entrepreneurial potential as well. New flexible and high level technologies were adopted by major industries and a new complementary service sector developed, creating a highly paid elite class (Keyder and Öncü 1993).

The Prefecture of Istanbul was founded in 1855. However, following the declaration of the Republic, municipal services were provided by the provincial government. In 1930, with the passing of the Municipality Law, number 1580, Istanbul Municipality was founded. In the following years changes in the government structure of Turkey also went into effect in Istanbul. As are other mayors in Turkey, the Mayor of Istanbul Metropolitan City is elected directly by the city's population for a five-year term and shares executive power with a Municipal Council formed by selected members of the city's 39 District Municipalities and their District Mayors. District Mayors are also directly elected and lead the District Municipalities. The Mayor of Istanbul Metropolitan City has extensive powers and a significant budget for city-wide planning, transport, housing and environmental services, amongst others (Burdett and Nowak 2009).

Due to the problems occurring as a result of rapid urbanisation, city-wide decisions regarding the concerns of citizens are crucially important in Istanbul. Local municipalities are not financially autonomous, so they are continuously searching for new resources in order to implement their programmes. On this point, good relations with central government are vital. In Istanbul, when the political party of the Istanbul Metropolitan Municipality (IMM) has differed from that of the central government, the most problems have occurred. This also applies to the relations between the IMM and the district municipalities.

Currently there is a two-tier structure in Istanbul's urban administration: both the metropolitan city and district municipalities have decision-making powers. The IMM is responsible for macro-level decisions concerning the entire city. District municipalities are responsible for decisions related to traditional municipal services (Erder 2009).

Urban governance in Istanbul is closely related to the election periods, especially in the last three decades, and the urban governance system is reflected in the large-scale urban projects. Especially starting in 1984 with the introduction of new legislation, the relative independence and the financial resources provided by the central government have strengthened the IMM. With law number 3030, a two-tier government system was introduced to metropolitan cities like Istanbul. In this system, there is the MM having control over the whole metropolitan zone and there are district municipalities within the metropolitan zone. As mentioned above, this relative independence granted to metropolitan city municipalities and financial resources provided by the central government have increased the controlling capacity of metropolitan city municipalities.

The first five-year period following the elections in 1984 marked the start of the *municipal mega-project* period in Istanbul. The new mayor of IMM was a member of the ruling party and the shift from managerialism to entrepreneurialism started and investments followed. Several large scale commercial and service projects took their place in the Istanbul development plan. Members of the Istanbul Metropolitan Municipality Council regarded the municipality as the investor, however, their main concern was urgent investment projects instead of macro-level long-run projects. Therefore, in this period, urban infrastructure and transformation projects such as the rehabilitation of the Golden Horn, the construction of the second bridge over the Bosphorus, and renewal projects prevailed. At the beginning of the 1980s, investments went into expanding the Central Business District, and a new axis, mainly with skyscrapers for offices and large-scale commercial projects, was also initiated. Although several improvements were achieved, criticism also existed. The importance of local politics increased and different groups in the city started to be involved in local politics in order to gain power to solve their own problems.

In the second five-year period, between 1989 and 1994, as a social democrat party member was elected Mayor of the IMM, the focus of investment shifted to the squatter housing areas. After 1994, in the following two election periods, members of the conservative party were again elected. In these two mandates, the focus was again on investment projects, but social service provision was also included, especially for low-income areas of the city (Erder and İncioğlu 2004, Uzun 2007). There were also new public–private partnership projects such as the Perpa Trade Centre Project. This represents an example showing how urban coalitions have been structured by informalities and conflicts with the government system.

The 2004 elections had important implications for urban governance and urban transformation projects. The new mayor of the IMM was again a member of the ruling party. This makes an important difference in the central and local government relations. Coordination between these two levels increased rapidly and the central government supported the metropolitan municipalities by enacting several laws that would facilitate governance and urban transformation. Some of the related laws are: Municipal Law (3 July 2005, law number 5393); the Law on protecting and restoring degraded historical and cultural heritage (16 June 2005, law number 5366); the Province Administration Law (22 February 2005, law number 5302) and MM Law (enacted on 10 July 2004, law number 5216). All of these laws steered new organisational structures towards urban governance. In addition, as the same mayor was elected in the elections held in 2009, the policies continued in the same vein.

At the moment, the metropolitan city administration of Istanbul follows the *powerful mayor and weak council* model. In fact, it is a wide and opaque space for macro-level decision-making. As a result, city-wide decisions are discussed and criticised in the media and by professional associations only after they have been made. Upon gaining power and all its political advantages, central government began toning down calls for decentralisation, and started exerting a strong influence in Istanbul. Moreover, due to the efficiency experienced at the local level, their decisions are generally supported by Istanbul's residents. In fact, especially in the last decade, Istanbul has been governed by a populist approach closely tied to the central government (Erder 2009).

Along with the partial projects, there are large scale projects designed for Istanbul by the IMM. In one such project, the aim is to make an analysis of the Istanbul Metropolitan Area and calculate the transformation potential of the city within the framework of planning Istanbul's future. Depending on this potential, a long-run vision will be determined. Besides the determination of strategies and policies, these are designed to enhance the global competitiveness of Istanbul and thereby locate the city regionally and globally. The most important point of this project is the cooperation between IMM, the OECD and the State Planning Organisation of Turkey. There are also other partners from the private sector, universities, and non-governmental organisations. The project also addresses sub-headings such as: socio-economic trends, advantages for competition and metropolitan governance (Uzun 2007). This was followed by an effort to improve coordination between the various departments of the IMM to help develop the city's master plan with the Mayor, also establishing the Istanbul Metropolitan Planning and Design Centre (IMPDC) in 2005. The centre was established with funding from a public–private partnership that serves as an affiliate company of the IMM. Though reduced in size in recent years, the organisation consisted of 400 experts, academics and key municipal members. They prepared the Istanbul Environment Plan at a 1/100,000 scale, which was approved in 2009 (Burdett and Nowak 2009). Although the staff of the IMPDC possessed expertise, decision-making was left to the domain of populist politics. IMPDC projects tend to be selectively or only partially implemented, as they lack administrative and technocratic influence. Technocrats are excluded from the decision-making process despite the fact that their knowledge and expertise are extremely important for the city's aesthetics and long-term growth (Erder 2009).

Another important project supported by central government was Istanbul becoming a European Capital of Culture. In 2005, a Council of Ministers decree was approved in order to establish an initiating project group for Istanbul 2010 European Capital of Culture. This group is composed of participants from the government, the municipality, non-governmental organisations and civil society members. After long processes and endeavours, Istanbul was deemed ready on 11 April 2006 to become the European Capital of Culture 2010 along with Pecs (Hungary) and Essen (Germany). On 13 November 2006, Istanbul was finally announced the European Capital of Culture for 2010 with the approval of the European Parliament and the European Union Council of Cultural Ministers. Following approval, the Istanbul 2010 European Capital of Culture (ECOC) Agency was founded in order to plan and manage preparation activities and coordinate the joint efforts of public bodies and institutions in order to realise this aim. The Agency operates in three strategic areas: culture and arts; urban applications and protecting cultural heritage; tourism and publicity. The official agency structuring was stipulated according to the Law on Istanbul 2010 European Capital of Culture enacted on 2 November 2007. According to this law, the Advisory Board comprises representatives from various ministries, the General Secretariat of European Union Affairs, Turkish Radio and Television, the Higher

General Directorate Board, General Directorate of Foundations, the Governorship of Istanbul and IMM together with 25 representatives nominated by the Istanbul Chamber of Commerce, Istanbul Chamber of Industry, Travel Agents Union of Turkey, the Istanbul Chamber of Architects, members of national and international bodies, those professional organisations which can be considered public bodies recognised for their activities in areas of culture, arts and tourism, and individuals renowned for their work on Istanbul's history, culture, architecture and art. There are also members from universities and opposition parties. On thc other hand, the Executive Board comprises one representative each from the Ministry of Culture and Tourism, the Governorship of Istanbul, IMM Istanbul Chamber of Commerce, Istanbul Chamber of Industry, two members selected by the Advisory Board from among civil society representatives included in the Initiative Group as indicated in provisional article 2, and two members of the Advisory Board selected from among other members of the Advisory Board.[3]

As can be seen from the composition of the Executive and Advisory boards, this represents an important public–private partnership for this high profile project. Therefore, this has been a very important project and experience for IMM as an example of governance. However, as the process proceeded, due to political choices and the restraints of the IMM, private sector participants involved in the process started to resign. These resignations changed the balance of the public–private structure in favour of public bodies.

Since 2004, another important issue for IMM has been urban transformation projects. Most projects targeting a rapid transformation of the squatter housing areas into modern residential areas see IMM working in cooperation with the Mass Housing Administration, which is a central government body. In parallel, public–private partnerships for urban renewal and urban development projects gained in importance. On the one hand, historical sites are being regenerated with projects receiving funding from foreign institutions such as UNESCO and the EU. Non-governmental organisations are also taking part in these projects. On the other hand, old industrial sites are scheduled for redevelopment. For instance, an international competition was held for proposals to transform an old industrial core into a new sub-centre, and the winning project is about to be implemented. In this example, the district municipality is also involved in the construction process. Another example is the construction of extensive shopping areas along with dwellings and cultural facilities where the IMM provides the necessary building permits and sometimes the land. In addition, such projects have both national and foreign partners as investors (Uzun 2007).

In addition, the governance of Istanbul does not happen only at the municipal and central levels. Partly to comply with the European Union accession process, Turkey recently created the Istanbul Development Agency, one of 26 regional

3 www.en.istanbul2010.org/ 2010AKBAJANSI/hakkında/index.htm (accessed 30 November 2009).

bodies assisting in coordination between municipal and central bodies as well as civic institutions for the budgeting and planning of large-scale urban projects.

As can be observed from the examples given, the IMM is always the controlling institution even while the project type and size may vary. The Municipality gains the financial and legislative support of central government due to their shared political backgrounds.

Conclusion

Urban governance is not a new phenomenon occurring as a result of globalisation. It is very obvious that with the effect of the globalising economy and subsequent changes, local government structures have changed. The process has shifted from governing to governance. Therefore, the number of actors involved in the decision-making and implementation stages of many public services have increased, with the private sector becoming widely involved. Together, these bottom-up policies have replaced some top-down policies. Many countries have been experiencing these changes and Turkey is to be counted among them.

Urban governance has been on the agenda since the 1980s for Turkey. Istanbul has been an example where some implications of these changes are observable. The transformation process still continues. At this point, it is not possible to reach a conclusion about whether it has been successful or not. On the other hand, the inner dynamics of different cities and societies must not be ignored.

General explanations may be applicable for explaining urban governance in Turkey. Nevertheless, local political cultures are still more effective in urban governance in Istanbul. Indeed, clientelism and patronage relations are always effective in urban governance. This culture has been closely related to electoral dominance over the past three decades. The political choices of mayors have had direct effects on the governance system, especially at the level of local and central government relations. It would, however, not be wrong to say that urban management systems are influenced by global forces, and the participation of the private sector in urban development has been increasing in Istanbul. Furthermore, the concept of governance does partially explain the urban management model in countries on the periphery, such as Turkey. Nevertheless, further comparative research must be made in order to ascertain the differences between different political and societal cultures.

References

Altaban, Ö. 1990. 1960–1973 Dönemi belediyeciliğine genel bakış, in *Türk Belediyeciliğinde 60 Yıl, Uluslararası sempozyumu, Ankara 23-24 Kasım 1990, Bildiri ve Tartışmalar*, Istanbul: IULA-EMME, 317–25.

Brenner, N. 2004. Urban governance and the production of new state spaces in Western Europe, 1600–2000. *Review of International Political Economy*, 11, 447–88.

Burdett, R. and Nowak, W. 2009. *Istanbul City of Intersections*. London: London School of Economics and Political Science.

Carley, M. 2000. Urban partnerships, governance and the regeneration of Britain's cities. *International Planning Studies*, 5, 273–97.

Davoudi, S. 1994. Dilemmas of urban governance, in *Managing Cities: The New Urban Context*, edited by P. Healey, S. Cameron, S. Davoudi, S. Graham and A. Madani-Pour. London: John Wiley & Sons Ltd., 225–30.

Erder, S. 2009. Local governance in Istanbul, in *Istanbul City of Intersections*, edited by R. Burdett and W. Nowak. London: London School of Economics and Political Science, 46.

Erder, S. and İncioğlu N. 2004. Yerel Politikanın Yükselişi, in *İlhan Tekeli için Armağan Yazılar*, edited by S. İlkin, O. Sillier and M. Güvenç. Istanbul: Tarih Vakfı Yurt Yayınları, 539–58.

Gülöksüz, Y. and Tekeli, İ. 1990. 1973–1980 Dönemi ve 1980 sonrası dönemi belediyeciliği, in *Türk Belediyeciliğinde 60 Yıl, Uluslararası sempozyumu, Ankara 23-24 Kasım 1990, Bildiri ve Tartışmalar*, Istanbul: IULA-EMME, 373–81.

Hamamcı, C. 1990. Tek parti dönemi belediyeciliği; Genel bakış, in *Türk Belediyeciliğinde 60 Yıl, Uluslararası sempozyumu, Ankara 23-24 Kasım 1990, Bildiri ve Tartışmalar*, Istanbul: IULA-EMME, 147–58.

Haus, M. and Heinelt, H. 2005. How to achieve governability at the local level? in *Urban Governance and Democracy*, edited by M. Haus and H. Heinelt. London: Routledge, 12–39.

Jessop, B. 1995. The regulation approach, governance and post-Fordism: alternative perspectives on economic and political change. *Economy and Society*, 24(3), 307–33.

Jessop, B. 1998. The rise of governance and the risks of failure: the case of economic development. *International Social Science Journal*, 155, 29–46.

Kepenek, Y. 1999. Türkiye'nin 1980 sonrası sanayileşme süreci, in *75 Yılda Çarklardan Chiplere*, edited by O. Baydar. Istanbul: Türkiye Ekonomik ve Toplumsal Tarih Vakfı, 229–40.

Keyder, Ç. and Öncü, A. 1993. Istanbul yol ayırımında. *Istanbul*, 17, 28–35.

Knox, P. and Pinch, S. 2000. *Urban Social Geography: An Introduction*. New Jersey: Prentice Hall.

Kovancı-Shehrin, P. 2005. *A Critical Evaluation of Governance in the Framework of Rural Development in Turkey*. Unpublished Doctoral Dissertation, Ankara: METU.

Marshall, T. 2000. Urban planning and governance: is there a Barcelona model? *International Planning Studies*, 5, 299–319.

Mayer, M. 1994. Urban governance in the post-Fordist city, in *Managing Cities: The New Urban Context*, edited by P. Healey, S. Cameron, S. Davoudi, S. Graham and A. Madani-Pour. London: John Wiley & Sons Ltd., 231–49.

Pierre, J. ed. 1998. *Partnerships in Urban Governance: European and American Experience*. New York: St. Martin's Press.

Rhodes, R.A.W. 1996. The new governance: governing without government. *Political Studies*, XLIV, 652–67.

Rosenau, J.N. 1995. Governance, order and change in world politics, in *Governance without Government: Order and Change in World Politics*, edited by J.R. Rosenau and E. Czempiel. New York: Cambridge University Press, 1–30.

Schimitter, P.C. 2002. Participatory governance arrangements: is there any reason to expect it will achieve sustainable and innovative policies in a multilevel context?, in *Participatory Governance: Political and Societal Implications*, edited by J. Grote and B. Gbikpi. Opladen: Leske+Budrich, 51–70.

Tekeli, İ. 1978. 1930–1944 Döneminde Cumhuriyet' in belediyecilik deneyimi, in *Türkiye de Belediyeciliğin Evrimi*, edited by E. Türkcan, Ankara: Ayyıldız Matbaası, 25–297.

Uzun, N. 2007. Globalization and urban governance in Istanbul. *Journal of Housing and the Built Environment*, 22, 127–38.

Chapter 11
Urban Governance in the South of Europe: Cultural Identities and Global Dilemmas[1]

João Seixas[2] and Abel Albet[3]

European Urban Governance

European cities have been positioning themselves in recent decades at a historical crossroads. The changes and restructures occurring in their rhythms, densities and landscapes, as well as in their broader to inner cognitive and political and cultural dimensions, have led European urban territories and societies into new types of opportunities and pressures. These opportunities and pressures are to be found in their most varied socio-political urban contexts, marked simultaneously by parallel confrontations in important features – from the continuous pace to absolute time-space flexibility and modularity of the economic and socio-cultural chains; to the crisis of the welfare state, raising new types of social needs and demands.

These fascinating but also disruptive times, in the conjunction of the heritage that François Ascher called the Fordist-Keynesian-Corbuosian paradigm (1995) with the development of hyper-territories, meta-cognitions and growingly complex functionalities of urban life, urban work, urban consumption and urban mobility, are framing new types of fluxes and externalities that clearly outstrip the capacities of the present political urban governments and institutional arrays.

At the same time, long-established socio-political structures and stakeholdings are also being reframed by these vast arrays of urban changes. What today seems widely recognised in most of the political, socio-cultural and academic realms is that this historical mutative scenario demands, from cities and urban societies, an absolute need to reinterpret several of their own structures and attitudes towards urban politics, urban administration and urban governance (Bagnasco and Le Galés 2000, Jouve and Booth 2004).

---

1 An early version of this text was published in 2010 in *Análise Social. Revista do Instituto de Ciências Sociais da Universidade de Lisboa*, 197, 771–87.

2 Instituto de Ciências Sociais; Universidade de Lisboa. Av. Professor Aníbal de Bettencourt, 9; 1600-189 Lisbon (Portugal). jseixas@ics.ul.pt.

3 Universitat Autònoma de Barcelona; Departament de Geografia. 08193 Bellaterra (Spain). abel.albet@uab.cat.

Multiple new urban-driven strategies policies were developed, many with promising (and realised) degrees of innovation and inclusion, others raising doubts about democratic procedures and cost-effective public deliverance. New types of urban projects and urban policies were consolidated; varied institutional structures were created; processes of administrative deconcentration and political decentralisation, some against the odds, were slowly raised; different arrays of principles and tools for urban strategy, urban planning, and even civic participation and civic rights, were proposed; more elaborated and influential forms of critical questioning upon urban socio-political regimes have been consolidated; political and instrumental improvements in social engagement and civic participation have been raised.

However, in spite of all these processes, the last two decades have also revealed certain blockades. Even for some of the seemingly most necessary political developments – such as the creation of metropolitan political authorities configuring stronger governance commitments at recognised scales of critical urban collective regulation and action; or the need for new both civic and public enforcement in face of possible deviation of democratic resources and procedures – many urban societies have been showing that the paces of their 'real cities' are not being adequately followed by corresponding paces on the part of their 'socio-political cities'.

This paradoxical scenario, having both wider opportunities for development and equity but also most challenging hurdles, seems to fully correspond to what Henri Lefébvre introduced 40 years ago as the long period of disorientation with the (then) expected outcome of the urban revolution (1970). This shows clearly to be the case for the present urban world of Southern Europe. Even – or especially – when in face of the dramatic economic and financial crisis that is particularly targeting its societies.

An important feature of this socio-political paradox seems to lie on a confusing conjunction of both classic regulatory strongholds with liberal demission attitudes on the part of the state towards the city. On one hand, recent decades have witnessed the gradual evolution of post-Fordist urban policies – and more recently even the reconfiguration itself of neoliberal urban policies – which tended to prioritise neo-Schumpeterian perspectives and to promote the enforcement of entrepreneurship and competitiveness (Brenner 2004, Harvey 2001, Jessop 1994). These perspectives were justified by expected provisions for the cities of higher levels of social, economic and creative qualifications, in a world of permanent appeal to new challenges in the areas of competitiveness, and sometimes of its own social emancipation and inclusion. But also, and relevant to our themes under consideration here, these were proposals also developed through the expectation of the enhancement of urban societies with a much stronger urban actor's activity, flexibility and proactive attitudes, thus catalysing governance networks and resulting in broader urban dynamics and socio-economic development.

On the other hand, however, severe criticism was raised about how it has been through these logics that structural changes have occurred in the political arenas and agendas, remodelling whole structures of urban politics and raising important questions regarding the potential deployment of main urban values such as equity, social justice and even democracy. For the critics, several years of neo-liberal dismantlement or even disruption of governmental public institutions have diffused (or fragmented, as some say) established political strategies and territories of public domain, these losing their prime role in urban provision and even in urban strategy and planning, with perverse repercussions on social and collective results.

Urban politics comprehends a vast arena where very different dimensions coexist, ranging from national strongholds to local political communities and to civic neighbourhood, from metropolitan strategic planning to human resources administration, from EU cohesion funding to real estate and swap finance. Within all these dimensions, the evolution of the forms of dialogue and conflict between different urban actors (between governmental and institutional organs themselves but obviously between these and the most varied actors of the civil society) remains a vast and triggering agora, full of lights and shadows. For thought as well as for action.

These are perspectives that follow the fields of the social sciences attentive to the city – which in truth should be mainly understood as a social construction – where emphasis is placed on the perceptions, identities, strategies and practices deployed in the actions of the multiple actors and communities living within the city's extent. This correspondingly recognises that *socio-cultural capital*, and what the literature refers to as *systems of action*, in a city are not bound only to specific urban design or urban planning configurations, but also incorporate the support structures and daily energies that leverage the city's destinies. The French sociologist Alain Touraine (1984) reflected on social life as a process, arguing for the necessary replacement of the *Society* concept by a *Social Life* concept, much more centred on the actors' actions and interactions:

> the essential is that the growing separation between the actor and the system might be substituted by its interdependency, by the idea of system of action … instead of describing the mechanisms of the social system, of its integration or disintegration, of its stability or change … we have to substitute the study of the social answers by the analysis of the mechanisms of auto-production of social life. (Touraine 1984: 31)

Through these scenarios filled with lights and shadows, an appreciation of the relational and processual concept of urban governance has been evolving. Let us therefore recall one of the most interesting definitions of governance: 'a process for the coordination of actors, of social groups and institutions in order to achieve collectively discussed and defined goals in a fragmented or even obscure environment' (Bagnasco and Le Galés 2000: 26). It is as if each city, and within it each project, policy or simple administrative process, should be seen as a collective construction whose success depends on the best or worst emergence of the interrelationship and co-responsibility networks, and the best or worst directions in the interconnection of its political, social and cultural forces, and the pressures and influences amongst the different actors on stage (Jouve 2003, Pinson 2009). These perspectives evidently have to be supported with the existence of a considerable degree of concrete rationality in governance management, thus implying the existence of dialogue and consensus-building structures across several scales: spaces, instruments and mechanisms, both formal and informal, through which conflict and cooperation fluxes might be processed with considerable proximity and the formation of interdependencies and partnerships is materialised with sizeable doses of objectivity.

Sizeable doses of democracy, inclusion and transparency are also called for, as the notion of urban governance, and its enduring and still quite appealing potential in the settings of urban politics, also carries important risks and mirror-side perspectives. This is so, firstly, because the simply utopian consideration of the city as a collective actor might bring (despite its virtues) obvious difficulties of consistency, often entailing concrete risks of reification – thereby amounting to nothing more than a constant deconstructivism; and secondly, because these discourses might also be the way for the consolidation, in many cities, of oligarchic governance decision-making political communities, through partial consensus-building processes, and thus not necessarily contributing to collective objectives. The fact is that after more than two decades in the spotlight of many academic and political debates and proposals, governance retains, and has even expanded, its light and shadows.

Political Dilemmas and Opportunities for the Cities of the South of Europe

Some authors – though not that many – have been examining the differences and specificities of the South European cities, namely in face of the most recent urban major challenges (see Borja 2003, Chorianopoulos 2002, Domingues 2006, Leontidou 1990, 1993, 2010, Nel·lo 2001). Most of these authors consider that for many of these southern cities and metropolises, there has been a distinctive path of urban development and restructuring, as well as distinctive modes of governance, at least throughout the major part of the twentieth century. Amongst other geopolitical and cultural specificities, this caused not only quite specific urban production processes (strongly understood in major trends like the peri-urbanisation of vast Mediterranean urban and coastal areas), and also sharp reductions in most cities qualification and competitiveness (Chorianopoulos 2002).

The differential focusing understood on these few scientific and critical analyses upon most recent socio-political developments in Southern European cities, when compared with most European urban scientific analysis, reinforce our view that the Mediterranean city *genius loci* still remains weakly analyzed by academia, if not misled. Following Leontidou (1990: 2), 'some of their everyday manifestations like informality, community life and socializing, song and football attendance, or mutual aid and illegal building, meet the indifference and scorn' of most of the academic and socio-political theorisers, fed on Marx, Weber and other major Northern/Protestant social thinkers, and 'are taken advantage of by the State' – this leading to a situation where 'creativity and spontaneity thus oscillate on the verge between opposition and cooptation'.

Our present purpose is to better understand corresponding differences, pluralities and common features structuring urban governance in cities like Athens, Marseille, Palermo, Barcelona, Ljubljana, Rome or Lisbon. As a differential social, cultural and geopolitical territory, Southern Europe has its own urban governance specificities that deserve attention. These specificities rest on social and cultural pillars, often impacting several institutional and governmental structures and all their normative and regulatory edifications (Seixas 2008). As both a legal and non-legal product of the socio-cultural stances existing in each urban society, urban governance (and its networks, stakeholdings, projections, democratic and non-democratic expressions) is a dimension that clearly reflects the existing specificities of every region. It also convenes and expresses the heterogeneities and dilemmas projected from every city into its own future.

Besides the role of the local scale, the nation-state still preserves a major role in defining each city's positioning and corresponding urban governance configurations and dynamics. This seems to be true even in such decentralised states as Spain, or in cities with a powerful economy like Paris or Milan. For many, 'La République contre la ville' as expressed by François Ascher (1998), showed and still shows to be the major framework where the national, regional and urban governance networks structure themselves. This is also the case even when, as still occurs in vast urban Mediterranean territories, the dominant role of the national governmental institutions proves to be distant and even dismissive regarding attentive forms of strategic urban planning and urban development policies, thereby arresting regulatory functions at the local level (Chorianopoulos 2002), and introducing distortions and gaps in the confrontation between urban needs and each city's governance capacities (Seixas 2006).

This generalised disruptive panorama developed different urban governance dynamics and embeddedness capabilities in different urban socio-cultural city configurations. Several cities have had difficulties in both their socio-cultural affirmation and their political representation at more effective influential scales. This has left fewer resources and capabilities for their local governments and societies towards their own city development. This seems to be the case for most of the Greek and Portuguese cities. Other cities, on the contrary, have not only managed to have direct or semi-direct political representation in important political arenas – such is the case of several French cities whose *maires* have been influential deputies on the Assemblée Nationale or even ministers in a long-established political tradition; or the case of some Spanish regional capitals, whose main leaders have had occasional political influence in Madrid. Some cities have also been able to develop local and regional networks of urban and socio-cultural governance, with corresponding results in their political and civic dynamics, and (obviously) with corresponding results in urban restructuring.

At the same time, due to their specific positioning in the local/regional to national confrontations, the capital cities of these South European nation-states seem to reflect in a higher form these different mirrors and properties. Their political capabilities range from the regional-configured considerable autonomy of Metropolitan Madrid (personifying in itself the double-face of Spanish politics and its constant tug of war between national and regional levels) to the severe trapping paradoxes of Athens and Lisbon (these, on the contrary, resulting with very little local and regional autonomy and highly fragmented governance panoramas, facing strong and ever-present national powers) and to the constant internal and politico-elitist struggles over Paris's effective government, raising the governance stakeholdings in the city of light (for ages, the main urban and cultural light for many Mediterraneans) as probably the best demonstration of the political battles inherent to Southern European cities.

In order to best develop and deliver public policies, and to better obtain the support of resources and stakeholders, several Southern Europe cities have gone in search of new types of strategies, processes, structures, and for more efficient solutions to manage and deliver public policies and to channel the dizzying transformations experienced in the city and its citizenry. Urban policies have been increasingly faced with the challenge of their own redesign being conditioned by the introduction of inter-institutional needs for cooperation between several governmental levels, and by parallel needs of deeper coordination between multiple agents and interests. For at least the last two decades, the options have often been to apply more liberal models in the construction and management of urban policies and projects – even when public coordinated, as would be expected for most of the South European cities. On the other side, however, many other strategies and policies maintained and even reinforced a considerable public and institutional strategic control over urban projects, their developments and results, even when including new governance forms and designs. Nonetheless, the tendency to follow one or the other political perspective seemed to depend more on the conjugation of the ambitions of each city's elites and main stakeholders – and the corresponding strategies and urban projects – with its own local and regional governance capacities and degrees of autonomy, than on supposedly more concrete ideological or partisan choices.

Presently, almost none of the main Mediterranean urban regions has a concrete metropolitan-scale government – with the relative exceptions of the Madrid Autonomous Community, designed to avoid an effective 'Federal District' for the city; and the Istanbul Metropolitan Municipality, with a certain degree of autonomy but constantly conflicting with central authorities. Some metro-governments existed but were abolished (as in Barcelona), while others may be considered as failed proposals (as the Italian cases). These administrative scenarios lead to enormous political gaps and huge practical problems, especially in crucial dimensions such as urban planning, public and private transportation, energy, and environmental issues. There are, however, public and functional-oriented metropolitan governance arrangements, but all too often subject to considerable doses of political communitarianism and lack of integrated coordination. These difficulties stem from all local levels and scales, resulting from inherited deficits or from present-day internal competitiveness between the different cities and towns that configure each metropolitan region. Most of the political and functional resources on the public administration apparels are driven to organically adapt to each specific metropolitan governance configuration – configuring highly complex governance structures in a panorama that most of the time has no clear basis of collective strategy and public delivery objectives.

That is why even in cities with a reasonable degree of autonomy and with recognised urban thought and strategy (as has been the case of Barcelona) the governance models followed, most visibly when materialised in large-scale projects – like the 22@ economic district or the Sagrera high-speed railway station in the Catalan main city, the recent major investments in Athens such as the Olympic infrastructures and the new airport, or the Expo and Parque das Nações major investments in Lisbon – faced severe criticism from many voices in their own urban societies (Albet 2004).

Surely the perspective of the urban *project* has been proving to be a relevant catalyser for urban governance dynamics, evolving economic and social agents, framing clearer objectives both to collective action and to local government administration (Pinson 2009), and when effectively local, also permitting some measure of civic participation and involvement. It has been through these projections that developing coalitions, political communities, and even pro-structural urban regimes of new types have been under rearrangement – their democratic culture and civic openness varying from quite plural forms to very closed, stratified and non-democratic choices. Local political agendas are, to a significant extent, today dominated by the logics of these project-driven regimes, often overshadowing other scales for political projection as well as local-type attentions, and leading the administrative frameworks to clearly prefer *new public management* attitudes to the detriment of *new public administration* actions (Mozzicafreddo et al. 2003), perceptively more complex to develop and surely much more delicate to negotiate in the present institutional, party political and labour union contexts. In reference to one of the main questions proposed by the French literature on these fields – *who governs the city* (Joana 2000) – although we do not consider that most of the urban regimes of major Southern European cities have evolved towards structured *glocalised competitive statist regimes* (as Brenner (2004) conceptualised for several urban regimes in the United States and Europe), we might consider that there is presently a considerable degree of power hypertrophy sustained through semi-closed political communities.

The cultural and symbolic identities that Mediterranean cities and their urban environments contain have become a crucial issue to attract new investments (related to tourism, creativity, heritage, knowledge, ICT, etc.) in order to ensure a greater competitiveness. Urban marketing and branding have been introduced as a regular promotional feature, although as a result, rapid processes of standardisation of urban projects and products have contributed to a relative banalisation of urban culture itself (Harvey and Smith 2005, Muñoz 2008). These transformations brought by this new cultural economy are also slowly reaching the huge peripheral territories of the Mediterranean sprawled cities – vast territories urbanised throughout the second half of the twentieth century and that clearly changed the historical geographical configuration of concentrated and dense cities, and where real estate and retail/distribution – and big public investments themselves – seem to continue to fragment social, economic and environmental resources in the name of enduring economic policies. And with the onset of the present severe economic crisis, tendencies might be for these coalitions and urban regimes to turn to more simplistic economic and symbolic competitiveness objectives, becoming more oligarchic and less participative, and leaving less space to the urban demands of large parts of the corresponding urban societies.

However, differential political and civic proactivity can also be noticed in several directions. With the existence of a wide and otherwise consolidated normative and political-institutional structure of government, there can be seen several areas of policy and administrative innovation, strategic thinking, and even democratic improvement. These processes bring perspectives for some change, together with other types of pressures and incentives deriving from newer origins (from the demands of the city-system and the urban society itself), but also pressures from other levels of government, namely the European Union, through administrative decentralisation enforcement, stronger local responsibilities, and new legal and fiscal frameworks. Altogether, these imply new demands, new attitudes and new positioning for urban governments.

The EU Factor

Southern European cities as a whole have proceeded to adopt proposals from the Central and North European world which seemed likely to provide more systematised and innovative solutions. On other occasions, 'home-grown experiments' were carried out successfully and retained for their original and beneficial solutions. Amongst the suggestions hailing from 'the other side of the Alps', it is worth remarking upon the policies drawn from the EU – such as the URBAN initiative – which have managed to play a noteworthy role in the introduction of new concepts and practices in urban areas of Southern Europe. However, the EU's urban policies have often failed to adjust to the social, cultural and financial idiosyncrasies of Southern European cities. The reverse also shows to be true: the singularities of Southern European urban policies have not always fit in with Northern European spatial priorities or with their neo-corporate intervention pattern, which is based on active policy integration of social and financial agents on a local level.

Overall, the innovations introduced by Southern European urban policies fall into two basic types. The first consists of new interactions between public institutions and key agents of civil society in an attempt to boost consensuses as well as social resources able to both formulate and implement urban policies: the key here must be found in the potential to produce public goods through social practices. On the other hand, there are all sorts of policies revolving around the interaction between public institutions and private businesses, the essential characteristic of which is the forging of partnerships between agents who share interests: novelty is rooted in the regulatory and administrative contexts that encapsulate these partnerships, therewith imposing both links and valuations.

One of the perhaps most direct consequences of the influence of the urban policies of the EU on new ways of city governance in Southern Europe is based on the spread of the competitiveness–cohesion dualism. This is why cities are considered as the driving force of economic growth, hubs for innovation, and key agents in the promotion and consolidation of international competition, as well as places in which various means of self-organisation are created as civic mechanisms devised to compensate for the deficiencies of markets and of the welfare state. This entails a sometimes exasperating duality that sways between the city understood as a hub for competitiveness (economic growth) and the city seen as a laboratory where new kinds of social cohesion and citizen welfare are fostered.

Those two concepts are usually deemed to be either mutually exclusive or symbiotic (cohesion in this case being a prerequisite to achieve competition). Many of the new types of governance thus speak of the need to boost competitiveness, meant exclusively in terms of structural transformation and urban economic growth, where the city is considered a collective agent who must capture resources that are scarce (such as economic investments, image, tourists, spectacular architecture) to secure an advantageous spot in the urban market.

On the contrary, as regards cohesion, it has essentially been seen as a formula intended to solve the many shortfalls and problems, legacy of the failures and dismantlement of the welfare state (the privatisation and externalisation of public services, increasing elderly population and growing foreign immigration, the difficulty of securing housing, etc.) – and far from seeing it as a social and solidary construction of inclusion and citizenship.

One of the major consequences of the implementation of EU policies might therefore reside in the strengthening of the belief that there is a close and consequent correlation between cohesion and competitiveness. This has led to the consolidation of a discourse according to which the success of a competitive city almost necessarily entails the widespread prosperity of the population. A major proportion of the investments accompanying these European initiatives have often significantly altered some urban landscapes, but not always taken into account the importance of cohesion as a social and spatial justice factor.

From many of the new proposals on urban policy recently drawn up in many Southern European cities, obvious changes have emerged in terms of the objectives and structure of public activity regarding the governance of the city and territory. Undoubtedly, much of this change has its origins in EU policy enforcements. But a further and quite separate issue is whether the new discourses on urban governance have meant effective changes in urban governing practices and the ability of all these policies to effectively address urban issues. Until now, in the Southern European urban world, it seems that in these matters there are more intentions than successes: the problems and difficulties that hinder changes and, above all, the power of inertia and inefficiency (resisting innovation in all its shapes) are all factors characterising the reduced level of application of new, genuinely transforming means of governance.

The new governance has yet to move beyond the conjunction of excessive rhetorical levels and good intentions, too often becoming embroiled in simplistic, superficial and manipulated debates regarding specific issues (such as security, immigration, etc.) and also beyond relatively closed political and bureaucratic communities often seeking pseudo-social approval for new policies and investments (such as airports, high-speed trains, major media events, etc.). It would therefore seem that for now (with some notable and praiseworthy exceptions), although new means of urban governance in Southern Europe show great potential in the realm of theory, these continue to present serious difficulties in terms of their actual implementation. As a result, criticism is occasionally voiced regarding several EU initiatives and, as a whole, the implementation guidelines of new models of governance which claim that these fail to reach beyond rhetorical attitudes, whilst no genuine, deep changes to urban policy are perceived.

So rather than just assessing the true parameters required to analyse the success and the effectiveness of *the* optimal urban governance, these ought to be based on the evaluation of the levels of appropriation and dispossession that the most different actors express. The reactions of citizens (which are almost always legitimate, though sometimes lacking in structure, apparently incoherent, or even not solidary in kind), are a strong mirror of the intricate fit between the interests of the administration and the civic expectations. Street demonstrations, associative life and general civic mobilisation processes are keys in the process of shaping and consolidating the rights that pave the way towards qualitative citizenship and 'the right to the city' for all.

Global Reflections upon Southern Urban Dilemmas

Vast arrays of new urban dynamics and pressures demand new urban policies – maybe a new overall urban politics. As stated at the beginning of this text, the diffusion and hyper-positioning of urban geographies and human daily realities is bringing a complete set of dilemmas and challenges to Southern European cities. Between socio-political reconfiguration tendencies and the risks of fragmentation of urban politics itself, governance and its capabilities of bringing together different and dispersed actors and aims surely opens new possibilities, but also new uncertainties. Are meridional urban societies quite aware of the pace of changes taking place, or do their cultural perceptions and socio-political structures remain at the side of contemporary risks and challenges? And are their respective structures and cultures of governance efficiently and democratically adapting to new realities and challenges, or do there exist significant imbalances causing limited and fragmented visions and political-administrative backwardness?

Surely, reinventing urban politics today means knowing how to better understand and construct collective action instruments, commitments and corresponding institutional management processes, able to better expand the human, cultural and relational wealth, thus improving social and civic capitals and generating clearer responsibilities upon collective problems (Subirats 2001). The perspective of each city as a local society (Bagnasco and Le Galés 2000) configured not only by formal but mostly by informal and organic governance networks, turns out to be highly relevant to the cities of the Mediterranean. Here lies one of its most (if not *the* most) triggering paradox: it is in the balancing between the strength of its socio-cultural complexities, the deep fragmentalities inherent to its spatial and political projections, and the (more recent) development of democratic principles and civic demands, that can be posed the higher potential to break with inertias and particularisms and to create interesting, innovative, and socially responsible cross-cutting socio-political proposals. As some researchers argue, although still with some vagueness and difficult objectivity, the fractalities of contemporary urbanity might ultimately well prove to be the most interesting metabolical base for new sorts of urban socio-political restructuring (Kooiman 2003, Rhodes 1997).

Vis-à-vis this complex background, still considerably opaque in the conceptualisation of its dynamics, it seems important to develop new types of questionings and to open new conceptual and analytical perspectives – both in the interpretation of the new political attitudes in the cities, and on their own capabilities to conveniently shape and project them – questionings that (in the domains of the academic research) have been raised in recent years through several fields of analysis, namely: the place and scope of city politics; the political economy and its urban projections; the urban governance debates; the urban regime approaches; the social capital and the cultural capital in the cities; the actor's strategies and the socio-political systems.

Recognisably, the governance focusing in the urban politics realms has been establishing new mechanisms and institutional procedures that can be driven both top-down and bottom-up. More autonomous and empowering agents appear in face of the traditional political parties and political institutions, as do policies needing to be developed through constant negotiation between the diverse agents and through consensuses built by a state that, in itself, also becomes more polycentric. In many South European societies, where the power and the role of the state has been traditionally strong but not so democratic and even less participative-driven, governance becomes an undoubted opportunity for different forms of inclusion and for diverse ways of political and public management. State, leaving the hierarchical, unidirectional and monopolistic structure of government, has a tendency to rethink 'down' its capabilities of government, and becomes just one more agent in a system of government more and more based on plural networks.

The somewhat different paces addressed in cities like Barcelona, Bologna and some French cities (notwithstanding some criticism of the relative variation of its outcomes), among many others, show nonetheless that urban governance networks can evolve through plural and democratic empowering manners, following effective pro-collective processes, objectives and public deliveries. Surely in these cities there exist specific characteristics that owe a great deal to considerable social capital directed towards urban and territorial self-development and autonomy (like the Catalan case), or to strong political enforcements and complex stakeholding governance networks (mostly the case of France). But these are precisely cases whose frameworks and dynamics should be better analysed and interpreted in the light of urban governance's possible developments.

It would be obviously too naïve to draw strict and overall generalisations when it comes to a territory as large and diverse as is the one that spans from Lisbon to Istanbul, crossing varied political realities from local to national and inclusively European scales. Some major frameworks on a specific southern city region – like, say, Rome, from its central state path-dependencies to specific configurations of its society's social and cultural capital – surely differ substantially from parallel frameworks for other meridional city regions like Marseille and Thessaloniki. The consequences on each city of the tendencies above expressed depend considerably on the potentialities, limitations, inner forces and dilemmas underlying the socio-cultural, political and administrative structures that exist in each city. This growing importance of the local and cultural spheres shows that it will very much depend on the urban socio-cultural capitals and stakeholding networks of each city and urban society, the resources of responsibility, and capability for the security and qualification of its own future. But following precisely this reasoning that puts culture as the most structural influential element for urban governance, one must at the same time give careful attention to the widest and most common cultural legacies affecting all these urban societies and territories – the Mediterranean culture.

Whatever the case may be, within the Southern European urban contexts here analysed several common features and respective interpretations and reflections stand out:

1. In most of the Southern European cities, the secular limitation of local powers when working towards decisive negotiative and resourceful capacities is remarkable. In many Mediterranean countries the weaknesses of local administrations, coupled with chronic issues regarding fiscal and financial support of its existence, have by and large conditioned their autonomy and political competences in terms of drawing up their own policies and thus local empowerment. On a number of occasions, the considerable weight of the central administration has failed to show an increase of infrastructures, equipment or services at the local scale, so that the welfare state has often been poorly (or belatedly) expressed on the urban Mediterranean scales. This lesser importance given to Mediterranean

cities compared with the higher local dynamism of Central and Northern European cities does not simply refer to the legacy of the industrial era (and consequently to the weaker economic and industrial growth displayed by many Southern cities), but has also been due to the spread of more organised and successful democratic processes and actors (including most influential forms of participation, cooperative activity, non-governmental activity) historically more inherent to the latter.

2. Over the last 30 years, some of these meridional countries have initiated regional decentralisation processes of varying scopes, which have brought about – with debatable success – a greater focus on intermediate and local territorial scales. These decentralisation processes still have not quite hit the right expression on the local level and have sometimes even proven detrimental. For one thing, it remains extremely improbable to develop political structures with effective power and that might better approach present main challenges (namely in the metropolitan and the micro-local or even citizenry scales). For another thing, cities still remain on the edges of the main developmental and political economy policies, despite their recognition as important developmental actors and the continuous rise of discourses considering the importance of higher local embeddedness. Nevertheless, on a few occasions some Southern European cities have stood out as references in urban socio-economic and political development in order to become something of a 'model' for other cities (like Barcelona, Bologna or Toulouse).
3. The considerable sprawl and socio-spatial fragmentation of the Mediterranean city – largely caused by meridional socio-political stakeholding structures and by the corresponding effects on the urban production models – seems to be, in the absence of effective metropolitan identity patterns, paradoxically fragmenting the traditional modes of urban governance and fomenting the loss of historical organic processes of local political stakeholding. This is an hypothesis not completely confirmed, though it is mostly based on the importance of the spatial and cultural bases for the configurations of meridional urban governance networks, from partisan to social and economic. But if confirmed, is this a tendency that might be exacerbating the political lags on administration, strategy and policy formation or, on the contrary, might it be contributing to configure new governance structures based more on territorial perspectives, and more directed to effective urban delivery issues? It seems that (also) one very important political direction to follow in the face of these disruptive tendencies should be to create regional-metropolitan institutions and governments – for these meta-governance formations and its socio-political resources to better objectify a large-scale political space to effectively influence the evolution of the entire urban region – thus being the basis for new forms of metropolitan governance networks.

4. Crucial doubts and uncertainties remain regarding local governance configurations and strategies – the idea and expression of governance not being by definition a guaranteed qualitative element in itself. This is strikingly true in urban societies like the Mediterranean ones, where social capital has always been complex and varied, but considerably fractal, highly personalised, or even populist, and not so much absolutely oriented to objective collective strategies or to effectively accountable democratic scenarios. And most notably when a quite dramatic financial and economic scenario of global austerity applies to most of these urban regions. In the last two to three decades, following discourses of strong catalysing projects, of cooperation and participation, of flexibility and optimisation of urban policies, urban governance has been too often understood in meridional Europe as a way for urban governments to stimulate populist or oligarchic regimes, or to discharge several of their responsibilities, often resulting in the disempowerment of strategic scales of action, in lower transparency, and weaker public control, in the avoidance of social objectives. And now, in front of severe pressures from the part of European political economy itself and the perspectives of deep curtails on welfare and public policies, these political oligarchic and urban disempowering tendencies might paradoxically be reinforced.
5. Surely influenced by European directives, for the first time national strategies of countries like Greece and Portugal have objectively recognised cities as a main asset for development and sustainability, thus raising the political and symbolic importance of their own urban territories. This is a tendency already consolidating for some time in regionalised Spain and in the quite territorially politicised France – precisely the two countries where the differential paces are particularly noticeable in urban policy and governance realms. This point – as the previous ones, actually – highlights the relevance of the state and its perspectives of political and administrative reorientation and restructuring, as a main actor precisely to permit (or cut back) the reinforcement of democratic metropolitan and local governance.
6. Finally, there are important novelties occurring in the socio-cultural and civic dimensions in Southern European urban societies. As Leontidou expressed (2010), there has been a notable cultural tendency in the Southern European civil societies, steadily observing the maturation of the cosmopolitanism of its inhabitants. These are transformations that can be understood through the widest social landscapes, from quite different lifestyles to the most varied urban social movements and civic demands, notably expanded by the dramatic spread of both economic austerity and political discontentment. This is an evolution that deconstructs the traditional North–South divide (and several other dualisms), but that at the same time 'broadens geographical imaginations in Europe' (Leontidou 2010: 1197). These urban civic expressions are rapidly moving towards much more sophisticated forms and contents, their development being

itself made through more organic-driven processes. Overall, it is a dynamic civic and cultural panorama that is certainly framing a new political culture and that will certainly have profound and long-term influences on the governance and political spheres of the Mediterranean cities.

References

Albet, A. 2004. La cultura en las estrategias de transformación social y urbanística de las ciudades. *Cidades, Comunidades e Territórios*, 9, 15–25.

Ascher, F. 1995. *Metapolis ou l'Avenir des Villes*. Paris: Éditions Odile Jacob.

Ascher, F. 1998. *La République contre la Ville. Essaie sur l'Avenir de la France Urbaine*. Paris: Éditions de l'Aube.

Bagnasco, A. and Le Galés, P. eds. 2000. *Cities in Contemporary Europe*. Cambridge: Cambridge University Press.

Borja, J. 2003. *La Ciudad Conquistada*. Madrid: Alianza Editorial.

Brenner, N. 2004. Urban governance and the production of new state spaces in Western Europe, 1960–2000. *Review of International Political Economy*, 11(3), 447–88.

Chorianopoulos, I. 2002. Urban reestructuring and governance: North-South differences in Europe and the EU urban initiatives. *Urban Studies*, 43, 2145–62.

Domingues, A. 2006. *Cidade e Democracia: 30 anos de Transformação Urbana em Portugal*. Lisbon: Editora Argumentum e Ordem dos Arquitectos.

Harvey, D. 2001. *Spaces of Hope*. Edinburgh: Edinburgh University Press.

Harvey, D. and Smith, N. 2005. *Capital Financiero, Propiedad Inmobiliaria y Cultura*. Barcelona: Museu d'Art Contemporani de Barcelona.

Jessop, B. 1994. Post-Fordism and the state, in *Post-Fordism: A Reader*, edited by A. Amin. Oxford: Blackwell, 251–79.

Joana, J. ed. 2000. Qui gouverne les villes? *Pôle Sud. Revue de Science Politique de l'Europe Méridionale*, 13, 3–11.

Jouve, B. 2003. *La Gouvernance Urbaine en Questions*. Paris: Elsevier.

Jouve, B. and Booth, P. 2004. *Démocraties Métropolitaines. Transformations de l'État et Politiques Urbaines au Canada, en France et en Grande-Bretagne*. Sainte-Foy: Presses de l'Université du Québec.

Kooiman, J. 2003. *Governing as Governance*. London: Sage Publications.

Lefébvre, H. 1970. *La Révolution Urbaine*. Paris: Gallimard.

Leontidou, L. 1990. *The Mediterranean City in Transition. Social Change and Urban Development*. Cambridge: Cambridge University Press.

Leontidou, L. 1993. Postmodernism and the city: Mediterranean versions. *Urban Studies*, 30(6), 949–65.

Leontidou, L. 2010. Urban social movements in 'weak' civil societies: the right to the city and cosmopolitan activism in Southern Europe. *Urban Studies*, 47(6), 1179–203.

Mozzicafreddo, J., Gomes, S. and Baptista, J. eds. 2003. *Ética e Administração. Como Modernizar os Serviços Públicos*. Oeiras: Celta Editora.

Muñoz, F. 2008. *Urbanalización. Paisajes Comunes, Lugares Globales*. Barcelona: Gustavo Gili.

Nel·lo, O. 2001. *Ciutat de Ciutats*. Barcelona: Empúries.

Pinson, G. 2009. *Gouverner la Ville par Project. Urbanisme et Gouvernance des Villes Européennes*. Paris: Presses de Sciences Po.

Rhodes, R.A.W. 1997. *Understanding Governance: Policy Networks, Governance, Reflexivity and Accountability*. Buckingham: Open University Press.

Seixas, J. 2006. A reinvenção da política na cidade. Perspectivas para a governação urbana. *Cidades, Comunidades e Territórios*, 12–13, 179–98.

Seixas, J. 2008. Dinámicas de gobernanza urbana y estructuras del capital socio-cultural en Lisboa. *Boletín de la Asociación de Geógrafos Españoles*, 46, 121–42.

Subirats, J. ed. 2001. *Elementos de Nueva Política*. Barcelona: Centre de Cultura Contemporània de Barcelona.

Touraine, A. 1984. *Le Retour de l'Acteur*. Paris: Maspero.

Index